Towers of Atlantis

Evidence not Fantasy

Paul Dunbavin

The right of Paul Dunbavin to be identified as the author of this work has been asserted herein in accordance with the Copyright, Designs and Patents Act 1988.

Whilst every effort has been made to identify and acknowledge the source of all quotations and illustrations it may be the case that some have been omitted and we apologise for these instances.

British Library Cataloguing in Publication Data
A catalogue record for this book is available from the British Library

© Paul Dunbavin and Third Millennium Publishing (2019)

ISBN: 9780952502937

CONTENTS

PREFACE

In 1995 I first published *The Atlantis Researches*. It was a result of my cross-disciplinary researches into pole shifts and how they might have affected world sea-level and ice-age climate. It was republished in 2003 as *Atlantis of the West*. The publisher preferred to stress the popular title in pursuit of sales as one would expect. The original sub-title: 'Britain's Drowned Megalithic Civilization' was also at the suggestion of the publisher – but then they dropped it anyway because it wasn't what the American co-publisher needed in order to gain sales in their market!

In this lies a more general problem: a book on a cross-disciplinary subject will find no single library classification under which it suitably fits. Does a study of ancient Atlantis and catastrophism sit under ancient history; or is it mythology? Certainly not archaeology! Neither can one put it out under science, such as climate or sea-level research. As soon as an academic referee sees the words they stop taking the subject seriously. Usually it has to sit under esoteric or 'mind, body, spirit' – but it isn't that either! What is esoteric about geophysics or the ancient climate? Fortunately, the freedom of internet searches will overcome some of the older constraints of classification – but academic researchers will still only look within their own specialist media.

The same problem occurs with this new volume too. The challenge is to say the same thing again as in 1995, without saying all the same things again – if that makes sense. A simpler summary of the evidence was required together with an update of the science and discoveries. In the intervening twenty-five years, nothing has arisen to negate the earlier research, indeed new discoveries have only added to its validity. In my earlier books I presented in an academic style and invited the reader to work out the cross-disciplinary pattern; but very few people will do that, certainly not reviewers – after all, they see so many books.

Joking aside, this is a very real problem as old as the 'arts versus science' debate itself. The experiments of a scientist can always be repeated and proven; equations can be checked; but a matter of mythology or history is merely the opinion of an eminent so-and-so, citing an earlier specialist, citing an even earlier one – right back to the very oldest sources that would never pass a modern referee. Archaeology (of which Egyptology is but a branch), climatology and glaciology sit half-way between, with field discoveries interpreted by the opinion of an expert. One cannot repeat evidence from the ground like a lab experiment. Astronomy and geophysics too, may fall into this halfway zone. So for a cross-disciplinary researcher to suggest something non-standard is always going invite knee-jerk responses from the deeply entrenched camps of specialist academia.

I hope you will find this book to be fairly light in tone so that readers who are only half-interested may dip in and out. There should also be some room for humour! For the most part it is a study of mythology and legends cross-checked against other disciplines. However, a presentation of real evidence cannot be structured like a work of fiction and that evidence may not always be exciting. I shall refer back to my earlier work rather than repeat the older references, preferring to cite new sources where needed and to pull on a few of the threads that were left hanging; for mythology is a subject where one simply cannot explore every interesting avenue.

This introduction is the only chapter where I shall use the first person singular. In *Atlantis of the West* I explored a theory of pole-shifts as a contributory cause of the ice-ages and climate/sea-level changes; and that led on to the subject of the Atlantis legend as only one of many sources. The pattern repeated here came out of that research – not the other way round; I was never directly seeking Atlantis or to publish a book about it. However, the subject here is most definitely the Atlantis legend and related mythology; and to follow the trail of evidence to find where it might lie. I stress: following the evidence.

In *Atlantis of the West* I pursued the likely astronomical causes and the characteristics of a pole shift; leading to worldwide episodes of climate and sea-level changes that have occurred since the ice-ages. In settling upon the most recent of these episodes around 5,000 years ago I then summarised a pattern of sea-level change along the Atlantic coast of Europe and of rising land further east. Out of this pattern there appeared a map of the former coastlines around Britain and Ireland that agreed surprisingly well with the description given by Plato. An analysis of the indigenous myths and legends of Britain and Ireland added further weight to that hypothesis. Modern evidence of sea-level and climate would (almost) support it too, but in this area there remains work to be done. In my later book *Under Ancient Skies* I explored the evidence of ancient astronomy and how that too can give us clues about a catastrophic past. I hope therefore with this forewarning of what is to follow, that you will pursue the sources yourself and come to your own conclusions.

Unfortunately there has been so-much nonsense heaped upon twaddle and put out by publishers on catastrophist subjects that it is difficult to know the best way to present the evidence. There are so many enthusiasts who have been drawn-in by this dubious earlier stuff that no-one can now come to the subject without scepticism. In responses to my earlier work: academics would assume that I was just another crank author; while the crackpots would presume that I was an academic – there is no middle ground. Publishers know that there are more enthusiasts than knowledgeable readers and so that is where they try to market their books. For a time there were quite crazy theories (some I am sure not even believed by their authors) that Atlantis had floated off to become Antarctica; and even a fictional TV series built on that one. I enjoyed the science fiction too, but it doesn't aid serious enquiry. Even to suggest in an academic paper that an ancient author might hold useful evidence for a modern geophysicist would evoke a referee's instant scorn.

The earlier books were published just before the internet came to replace published paper as a primary research tool. In reviews of my earlier work I would read: Paul Dunbavin conjectures this; he speculates that. *No he does not!* He presents a pattern of cross-disciplinary evidence and then summarises it. It is *the pattern of evidence* and the *coincidences* that suggest where we should seek the hard evidence of an ancient catastrophe and where a submerged city might lie; and if it be not there then those coincidences will remain and they will need to be explained in some other way. So here, I shall place more stress upon the coincidences and the 'mythological fossils' that suggest the pattern.

To define this idea of a mythological fossil: when a historical event loses its evidence of date then it becomes a timeless 'legend'. Similar sounding events and characters may then become confused and jumbled out of sequence in the retelling or in translation to another language. A legend may degrade still further into a 'myth' with the events perhaps associated with deities or stories of origin and creation that form part of a religion. Hence, to say that something is mythical has become the same as to say that it is not true! By this progression is the most ancient history lost to us.

However, significant details may retain their integrity through his process and we may extract and examine them with modern science, or compare them with other evidence. These are the 'mythological fossils'. We should look for precise and seemingly irrelevant details that would not strictly be needed in a fictional plot. These give us a clue that the story is authentic. And even a fictitious plot, as any author will tell you, cannot be constructed without a backdrop of contemporary real geography and politics. This background too may point us to the likely date of such events.

In schooldays when I learned chemistry I was taught to write-up my science as: object – method – results – conclusion. My *object* here is to examine the myths scientifically; the *method* is to isolate elements in the myths that can be checked by science or against other sources. The *results* are then written-up and summarised with a *conclusion*. This is not speculation.

In an era of instant information via internet-searches, blogs and fake-news it seems that no-one wants to do the detailed research any more. Anyone can check a fact simply by putting it into a search-engine. No-one wants to trek round libraries looking for old out-of-print books as I had to do in the 1980's. They want to go straight to the answers and then seek some 'expert' to confirm: 'yes, this is good, this is right'; but the reality is that for a cross-disciplinary subject there are no such multidisciplinary experts. I bow to the knowledge of any academic in their field of excellence – but once trap them outside their safe zone and you may as well seek the opinion of a bus driver. Will a professor of Greek understand the equations? Does the physicist give a hoot for ancient myths?

The archaeologist Stuart Piggot notoriously once said that only archaeologists should be allowed to put forward theories about the past. That would be rather like suggesting that stamp collectors should best run the postal service because they know a lot about stamps. On the other side of that coin we may find field researchers, who think they see changes in astronomical alignments of monuments, feeling obliged to cite out-of-date Velikovsky theories in order to have their papers accepted. Between these two camps there lies a gulf as deep as the Atlantic itself. Theorists need the field researchers; we should cite them, not the other way about. I will always respect the painstaking work of those specialists who dig in marshes and fields, if only they would in turn value the work of the cross-disciplinary researcher. All we can really do is to survey the wood from the outside while the specialists examine the trees and say to them: 'here's a good place to take a closer look'.

Therefore, if ever you should find a disparity between an ancient source and the opinion of a modern scholar then *always* trust the ancient source. A historian will invariably prefer the account of an eye witness, or the person closest to the events, over an account that is written down years later. Why then do we do the opposite when investigating prehistory? Show some respect for the ancient stories. Probe them with science.

PREFACE

1

Ancient Sources – Ancient Truths?

Plato is one of many ancient writers who offer us evidence of catastrophic events in recent prehistory; but other Greek, Roman and Byzantine historians had access to older philosophical works that are now lost to us. Most useful for our purpose are Diodorus of Sicily (80 BC-20 BC) and the Neo-Platonist philosopher Proclus (AD 412-485). Their works are summarised here with the most significant passages quoted.[1]

The *Timaeus* of Plato introduces the submergence of an island called Atlantis as part of a dialogue between four philosophers: Timaeus himself, Socrates, Critias and Hermocrates. It is believed to be one of Plato's final works before the *Critias* and *The Laws*, which was left as unfinished on his death in 348 BC. It is important to appreciate that neither Atlantis, nor the various catastrophes that have affected the Earth, were the principle motivation for the work. The *Timaeus* continues with a wider discussion of science and astronomy together with the organisation of the world and the nature of the gods. It tells us that the *Critias,* containing the description of the island, is to follow; and a book by Hermocrates (which is lost or never written). The real purpose of all this was to contrast the political organisation of the ancient island with the idealised society previously proposed by Socrates. Classicists will therefore declare that it is unimportant whether Plato believed the Atlantis myth to be history or fiction - merely that it represented a form of governance with which to contrast the democracy of Athens and the Greek city-states.

Critias, as speaker, relates a story of the achievements of the earliest Athenians as bought back from Egypt by Solon, one of the Seven Sages of Athens and a close friend of his great-grandfather Dropides; the story had then been passed down to him.[2] In the *Critias* narrative he records;

> My grandfather had the original writing, which is still in my possession, and was carefully studied by me when I was a child.[3]

The story told how the ancient inhabitants of Athens, the Hellenes, had repelled an attack by a great empire whose base lay along the coasts of the Atlantic Ocean. Solon visited the city of Saïs in the Nile Delta and had spoken with the priests of the goddess Neit. According to these priests, she was the same goddess as the Greek Athene.

An elderly Egyptian priest tells Solon that there have been many cataclysms of fire and water in human history, but that Greeks remember just one – the flood of Deucalion. He recalls the Greek myth of Phaeton, which was also known to the Egyptians:

> Now this has the form of a myth, but really signifies a declination of the bodies moving in the heavens around the earth, and a great conflagration of things upon the earth, which recurs after long intervals; at such times those who live upon the mountains and in dry and lofty places are more liable to destruction than those who dwell by rivers or on the sea-shore. And from this calamity we are preserved by the liberation of the Nile, who is our never-failing saviour.

Viewed with the knowledge of twenty-first century astronomy we should regard this observation as surprisingly prescient, from whomever it originates. We have recently observed the impact of a comet on the planet Jupiter and it is now generally accepted that such impacts on Earth have caused mass extinctions of species throughout prehistory.[4] We have seen the destructions caused by tsunamis in Sumatra and Japan and must assume that the Egyptians recorded in their annals the effects of the Santorini (Thera) eruption on their coastline as well as lesser events.

The priest relates that everywhere people survive these events but that history was preserved in Egypt because they alone possessed writing and recorded it in their sacred annals:

> And whatever happened either in your country or in ours, or in any other region of which we are informed...have all been written down by us of old, and are preserved in our temples.

He goes on to describe to Solon the ancient catastrophe. Before the great Deluge, there existed in Athens the finest race of men, capable of the noblest deeds in war and peace as taught by the goddess;

> She founded your city a thousand years before ours...and afterwards she founded ours, of which the constitution is recorded in our sacred registers to be 8,000 years old. As touching your citizens of 9,000 years ago... the exact particulars of the whole we will hereafter go through at our leisure in the sacred registers themselves.

He elaborates that the laws of Athens and Egypt are almost identical, because they had the same origin, but the greatest of their deeds lay in the defeat of the invaders from the Atlantic.

> For these histories tell of a mighty power which unprovoked made an expedition against the whole of Europe and Asia, and to which your city put an end. This power came forth out of the Atlantic Ocean, for in those days the Atlantic was navigable; and there was an island situated in front of the straits which are by you called the Pillars of Heracles; the island was larger than Libya and Asia put together, and was the way to other islands, and from these you might pass to the whole of the opposite continent which surrounded the true ocean; for this sea which is within the Straits of Heracles is only a harbour, having a narrow entrance, but that other is a real sea, and the land surrounding it on every side may be most truly called a boundless continent. Now in this island of Atlantis there was a great and wonderful empire which had rule over the whole island and several others, and over parts

of the continent, and furthermore, the men of Atlantis had subjected the parts of Libya within the columns of Heracles as far as Egypt, and of Europe as far as Tyrrhenia. This vast power, gathered into one, endeavoured to subdue at a blow our country and yours and the whole of the region within the straits; and then, Solon, your country shone forth...being compelled to stand alone...she defeated and triumphed over the invaders, and preserved from slavery those who were not yet subjugated, and generously liberated all the rest of us who dwell within the pillars.

At this point Critias describes the geological catastrophe to which he has alluded earlier in his narrative:

But afterwards there occurred violent earthquakes and floods; and in a single day and night of misfortune all your warlike men in a body sank into the earth, and the island of Atlantis in like manner disappeared in the depths of the sea. For which reason the sea in those parts is impassable and impenetrable, because there is a shoal of mud in the way; and this was caused by the subsidence of the island.

There is much more to the *Timaeus,* to which we may later return. To modern ears the geography does not seem surprising because we know it to be quite accurate. It is important to appreciate that this was not so for scholars of Solon or Plato's era. We need only read Herodotus, dating from around 450 BC, to gain an insight into how little was known of the geography of Western Europe by classical Greek and Roman writers. The interior of Europe was to them a temperate jungle inhabited by uncivilised tribes, as impenetrable to travellers from the Mediterranean as Amazonia would be to later explorers. Even the coasts of the Atlantic were mysterious to them.

Solon's visit to Saïs is independently verified by Plutarch (AD 45 – AD 120) in his *Life of Solon*; he supplies the name of the Egyptian priest whose evidence we follow. Solon was revered by later Greeks as one of the seven sages who established Athenian democracy, before retiring and leaving the implementation of his new laws to others. He was a hard-headed politician living in the dangerous world of Classical Athens – certainly not a man for flights of fancy. We are told that Solon spent as much as ten years on his travels and remained for a long time in Egypt discussing philosophy with *Sonchis of Saïs* and *Psenophis of Heliopolis*. This would have been around 590 BC some two hundred years before Plato.

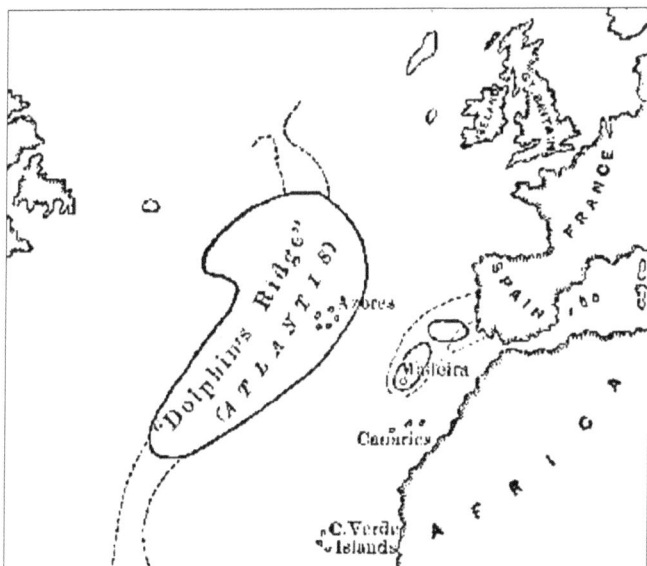

Figure. 1.1 A popular representation of Atlantis taken from Ignatius Donnelly's 1882 book 'Atlantis: the Antediluvian World'. Note it and then disregard it! The 'mid-Atlantic' location has served to pervert all subsequent analysis of the evidence; the real islands in the Atlantic are shown further north.

To return to the narrative of *Timaeus*, we may see that the invaders came 'from out of the Atlantic Ocean'. He does not however specifically state that they came from the island that he says lay 'in front of', or 'opposite' the pillars - or the straight of Gibraltar as we now know it. We must allow a little inexactness in translation and retelling by all the sources that bring the story to modern eyes. The description of an island, which was the way to other islands beyond and to the continent surrounding the ocean, is certainly captivating. To imagine that ancient explorers might have known of the American continent is one of the reasons why discussion of Plato's narrative has taken on a life of its own. But it is only the precision of that word 'opposite' that has led later commentators to place the island or 'continent' out in mid-Ocean.

The specific wording would infer that the great earthquakes occurred not long after the war in which the Atlantic invaders were defeated, as they enveloped all the soldiers. Again, we must allow some license. Earlier the priest implies that as much as a thousand years had passed between these events. We can only really rely on such details where there is corroborating evidence.

Diodorus and the Amazons of Libya

The Greek writer Diodorus Siculus has independently preserved for us many other mythological sources that may give glimpses of ancient history not found elsewhere. Diodorus is not spoken-of kindly by classical scholars, perhaps because his version of the ancient past was derived from unfashionable Egyptian and North African (perhaps Carthaginian) myths and legends, rather than the classical Greek historians and mythographers. Like all myths, they are timeless. Firstly we need to *trust,* rather than *dismiss*, their content and then seek the internal clues or 'fossils' that will allow us to align them with other available evidence.

Although Diodorus sought to treat the myths as a source of history, it is still not easy to arrange his rationalised histories into a clear sequence. He also describes a people called *Atlantians* who lived along the coast of the Atlantic Ocean. No one really knows how the name 'Atlantic' for the western ocean came about except that it is found in our earliest Greek and Roman sources, but we still glibly use the name today without a moment's thought.

Diodorus Siculus did not cite Plato as a source. We cannot be sure if he knew the Atlantis story in its entirety, however, he did know of the association of Saïs and Athens, for he confirms the Egyptians belief that it was it was originally a colony from Saïs and that evidence of this existed in their records.[5] As to why he did not cite Plato's narratives we can only rely on his own words:

> Now as for the stories invented by Herodotus and certain writers on Egyptian affairs, who deliberately preferred to the truth the telling of marvellous tales and the invention of myths for the delectation of their readers, these we shall omit and we shall set forth only what appears in the written record of the priests of Egypt and has passed our careful scrutiny.[6]

One wonders to whom he could have been referring in such a disparaging way! Little did he know that later scholars would denounce his own efforts in like manner. Still later, he observes:

> For the priests of Egypt recount from the records of their sacred books that they were visited in ancient times by Orpheus, Musaeus, Melampus and Daedalus, also by the poet Homer and Lycurgus of Sparta, later by Solon of Athens and the philosopher Plato...

We must presume therefore that the original records must have been first written in the primitive hieroglyphs that Egyptologists find on the earliest artefacts from Egypt; and then transcribed numerous times by later scribes, with all the potential for scribal error that comes with that. Certainly, the records existed; probably they were lost only when the temples fell out of use in Roman times. Diodorus encountered, either directly or through his sources, a similar long chronology of thousands of years as we find in Plato's narratives and this congruence lends authenticity to both.

The earliest 'history' that survives from Diodorus in his *Library of History* is unique, but offers tantalising details that may hold a grain of truth. It is rather like comparing Geoffrey of Monmouth's jumbled *History of the Kings of Britain* with verifiable history – real history keeps popping-up. Diodorus paraphrases the account of Dionysius 'Skytobrachion' (leather-armed) who lived in Alexandria and who is known to have rationalised many myths; he presumably had access to the lost books in the Great Library there; and we must accept the claim of Diodorus above that he personally checked everything against the Egyptian sacred records.

There was formerly in Western Libya a race of Amazons: that is, a tribe ruled by women; and whose young women were required to serve in the army and preserve their virginity.[7] At least, the Greeks called them Amazons ('without a breast') but what name they may have called themselves is unknown. Their men, he says, took no part in the military campaigns – perhaps difficult for us to believe today – but we are told that they fought mainly with cavalry and bow-and-arrow at that era, before metal swords and close combat.

The original home of the Amazons, as derived from their own myths, was on an island called *Hespera*: 'west'. It was surrounded by a low-lying area called the marsh Tritonis;

> This marsh was near the ocean which surrounds the earth...and that mountain by the shore of the ocean...and is called by the Greeks Atlas.

We can scarcely work out where this island and Mount Atlas were located. We are told that the land was full of fruit trees and flocks of goats and sheep, but they knew nothing of grain or farming. The fact that it also suffered from eruptions of fire has often led to it being identified with Diodorus' own home island of Sicily – but that island does not lie west of Libya. We are told that they conquered their own island and then subjugated the tribes of the marsh Tritonis, where they founded a city called Cherronesus, meaning: 'peninsula'.

At some later time, their queen Myrina gathered an army of thirty-thousand infantry and three thousand cavalry;[8] and they invaded the territory of the *Atlantians*. Now of these people we are told:

> The most civilized men among the inhabitants of these regions, who dwelt in a prosperous country and possessed great cities; it was among them, we are told that mythology places the birth of the gods, in the regions which lie along the shore of the ocean.

Diodorus goes on to describe how the Atlantians were swiftly defeated and their city named Cerné was conquered with great massacre; but it seems that later the Atlantians saw these women warriors as potential allies and persuaded them to attack their own enemies the Gorgons, who lived nearby in Libya.[9] We can only infer that these people lived somewhere between the Atlantic coast of Africa and the Nile; the geography is difficult because, as we shall see, there have been many coastal changes and transition to desert conditions since all of these supposed events. The defeated Gorgons were scattered into the forests for refuge after much slaughter of prisoners; however, we need not dwell any further on the Gorgons.

More importantly, we are told of Queen Myrina's subsequent conquests towards the east:

> As for Myrina...she visited the larger part of Libya and passing over into Egypt she struck a treaty of friendship with Horus the son of Isis, who was king of Egypt at that time. And then after making war...upon the Arabians... she subdued Syria...She also conquered...the races in the region of the Taurus...and descended through Greater Phrygia to the [Mediterranean] sea.

Therefore, we may see in all this an empire almost rivalling the later conquests of Alexander in extent, and leaving her near Pergamum, just across the Aegean Sea from Greece. Like Alexander, we are told that the queen then went on to found many cities, seizing the island of Lesbos and founding the city of Mitylene, named after her sister.[10]

The value of this tale about the Amazons is that it allows us more potential dating evidence for the era of the *Atlantians*. In arranging a treaty 'with Horus the son of Isis' it is usual to see in this the mythical god of the same name and so dismiss the entire story of Myrina and her Amazons to the timeless realm of mythology; but in fact all Egyptian rulers right up to the Ptolomies would describe themselves this way. Every pharaoh would become the new Horus. We may perhaps see in this a real dynastic or predynastic king of Egypt whom we have the opportunity to identify.

Figure 1.2 The Campaigns of Queen Myrina showing some of the places described by Diodorus. The location of the Amazons' homeland and the location of the conquered *Atlantians* cannot be determined, but we may presume that they lay somewhere between modern Tunisia and the Strait of Gibraltar.

Of the eventual fate of Myrina we are further told that in pursuing the conquest of these Aegean islands, she was caught in a storm at sea and took refuge on a then-uninhabited island, which Diodorus identifies as Samothrace (Samos): 'sacred island'.

> ...and she [the mother of the gods] established the mysteries which are now celebrated on the island...that the sacred area should enjoy the right of sanctuary.

We are told that shortly afterwards Mopsus, a Thracian, invaded the Amazons conquered territory and in the ensuing battle Queen Myrina was killed and the greater part of her army was lost. The survivors retreated to Libya and the Amazons' campaign of conquest was at an end.

Diodorus moves on to describe and rationalise the origin of the Greek gods among the Atlantians, but regarding the marsh Tritonis, he says:

> The story is also told that the Marsh Tritonis disappeared from sight in the course of an earthquake, when those parts of it which lay towards the ocean were torn asunder.[11]

This is another synchronism that offers potential dating evidence. It explains why we can no longer identify the geography and echoes the cataclysm of Plato. We cannot be sure, (as it is written) whether this earthquake occurred after the campaigns of Myrina, or whether it was earlier and prompted her quest for new territories. We cannot be confident that it was the same geological event as Plato describes or whether it occurred much later. After all, the loss of old Alexandria to the sea following earthquakes and tsunamis is well attested in historical times; and so earlier episodes should not surprise us. The marsh Tritonis is traditionally identified with the lake called Chott Djerid in southern Tunisia. Where the Atlantian city of Cerné lay remains unknown, but we must assume that it was one of the colonies mentioned in Plato's narratives.

The Critias – more about Atlantis

To return again to Plato, the remainder of his description of Atlantis is given in the *Critias*.

> Let me begin by observing first of all, that nine thousand was the sum of years which had elapsed since the war which was said to have taken place between those who dwelt outside the Pillars of Heracles and all who dwelt within them; this war I am going to describe. Of the combatants on the one side, the city of Athens was reported to have been the leader and to have fought out the war; the combatants on the other side were commanded by the kings of Atlantis, which, as I have said, once existed greater in extent than Libya and Asia, and afterwards when sunk by an earthquake, became an impassable barrier of mud to those voyagers from hence who attempt to cross the ocean which lies beyond.

The narrative continues with the description of the ancient Hellenes and the Acropolis before the earthquakes left the relict that we see today. In many places, he claimed, the land around the neighbouring Greek islands had also fallen away and been submerged.

Plato first warns us not to be surprised that Solon had translated the meaning of all the proper names of gods and heroes into Greek, intending to use them in his poetry. This is rather unfortunate for our modern enquiries and it has helped those who prefer to see it as Plato's own fiction. Many of these gods we may recognise as the Titans of Greek myth as given by the mythographers.

We are told that in the earliest times, the world was divided between the gods and Poseidon received the island of Atlantis. Poseidon and his children by a mortal woman settled in the island. It is important to quote the geography:

> Towards the sea, half-way down the length of the whole island, there was a plain which is said to have been the fairest of all plains and very fertile. Near the plain again, and also in the centre of the island at a distance of about fifty stadia, there was a mountain not very high on any side.

Poseidon is said to have settled there with his wife Cleito and to have encircled this mountain with rings of water, which over generations would become a great palace with springs of hot and cold water.

> ...making alternate zones of sea and land larger and smaller, encircling one another; there were two of land and three of water, which he turned as with a lathe, each having its circumference equidistant every way from the centre, so that no man could get to the island, for ships and voyages were not as yet.

The narrative describes how Poseidon would father five pairs of male twins and so divided the island into ten portions, assigning each to one of his sons as princes, with his eldest son given authority over all his brothers; and we have all their names (as we are told that Solon had translated their meaning to Greek). However, there are also useful geographical details to note in this description:

> ...the eldest, who was the first king, he named Atlas, and after him the whole island and the ocean were called Atlantic. To his twin brother, who was born after him, and obtained as his lot the extremity of the island towards the pillars of Heracles, facing the country which is now called the region of Gades in that part of the world, he gave the name which in the Hellenic language is Eumelus, in the language of the country which is named after him, Gadeirus.

These children and their descendants over many years would go on to rule not just over the island, but also the regions of the Atlantic coast adjacent. We may see in all this a myth of origin as we find in most cultures, which must already have been old and degraded history when it was recorded by the Egyptian priests. Again, note that it would not really be necessary, in a fictional account, to give such precise geographical detail. We now know that the island had a peninsula. Gades and Gadeira we can identify as the modern city of Cadiz in southern Spain and the province around it. It is this and the statement that the island lay 'opposite' the straits which would mislead later commentators to seek a sunken continent out in the open Ocean.

> ...and also, as has been already said, they held sway in our direction over the country within the pillars as far as Egypt and Tyrrhenia.

Now this statement is imprecise. We do not know whether the phrase: 'as far as Egypt' includes part of Egypt or if it stopped at its borders. We are elsewhere told that Horus was the last of the gods to rule over Egypt and that the home of the gods was among the Atlanteans. So make of this what you will. The reference to Tyrrhenia here specifically implies northern Italy and the flourishing Etruscan civilisation as existed in Plato's era.

Their island provided the kings with great wealth and again we are given specific details about their resources; metals, wood and wild animals, elephants and horses; and agriculture of fruits and herbs, yet without specific mention of wheat or grain.

> ... In the first place, they dug out of the earth whatever was to be found there...and that which is now only a name and was then something more than a name, orichalcum, was dug out of the earth in many parts of the island, being more precious in those days than anything except gold...Also whatever fragrant things there now are in the earth, whether roots, or herbage, or woods, or essences which distil from fruit and flower, grew and thrived in that land.

The Narrative moves on to detail the central island and the palace of Atlas. Now this is for many people the most compelling section and is again full of detail that goes beyond the needs of mere fiction:

> First of all they bridged over the zones of sea which surrounded the ancient metropolis, making a road to and from the royal palace. And at the very beginning they built the palace in the habitation of the god and of their ancestors, which they continued to ornament in successive generations, every king surpassing the one who went before him to the utmost of his power, until they made the building a marvel to behold for size and for beauty. And beginning from the sea they bored a canal of three hundred feet in width and one hundred feet in depth and fifty stadia in length, which they carried through to the outermost zone, making a passage from the sea up to this, which became a harbour, and leaving an opening sufficient to enable the largest vessels to find ingress. Moreover, they divided at the bridges the zones of land which parted the zones of sea, leaving room for a single trireme to pass out of one zone into another, and they covered over the channels so as to leave a way underneath for the ships; for the banks were raised considerably above the water. Now the largest of the zones into which a passage was cut from the sea was three stadia in breadth, and the zone of land which came next of equal breadth; but the next two zones, the one of water, the other of land, were two stadia, and the one which surrounded the central island was a stadium only in width. The island in which the palace was situated had a diameter of five

stadia. All this including the zone and the bridge, which was the sixth part of a stadium in width, they surrounded by a stone wall on every side, placing towers and gates on the bridges where the sea passed in.

Even geological details are supplied, which may help us in identifying location. We should ask again, why such exact detail would be needed in a purely fictional discussion of Greek politics,

The stone which was used in the work they quarried from underneath the centre island, and from underneath the zones, on the outer as well as the inner side. One kind was white, another black, and a third red, and as they quarried, they at the same time hollowed out docks double within, having roofs formed out of the native rock. Some of their buildings were simple, but in others they put together different stones, varying the colour to please the eye, and to be a natural source of delight. The entire circuit of the wall, which went round the outermost zone, they covered with a coating of brass, and the circuit of the next wall they coated with tin, and the third, which encompassed the citadel, flashed with the red light of orichalcum.

It may be seen that the rocks are not drab grey volcanic and seem to be layers of different sediments. Metals are mentioned, so again, we may seek places where both copper and tin are found in the earth. The precious orichalcum is problematic. Some commentators make this red gold (gold with copper); others prefer amber, but it is discussed along with metals. Knowledge of such metals at the earliest period of the Egyptian state would be precocious indeed, if detail of such metals (supposedly unknown to the early Egyptians) were recorded in their sacred records. Next the dimensions and features of the royal palace are described:

The palaces in the interior of the citadel were constructed on this wise:- In the centre was a holy temple dedicated to Cleito and Poseidon, which remained inaccessible, and was surrounded by an enclosure of gold; this was the spot where the family of the ten princes was conceived and saw the light, and thither the people annually brought the fruits of the earth in their season from all the ten portions, to be an offering to each of the ten. Here was Poseidon's own temple which was a stadium in length, and half a stadium in width, and of a proportionate height, having a strange barbaric appearance. All the outside of the temple, with the exception of the pinnacles, they covered with silver, and the pinnacles with gold. In the interior of the temple the roof was of ivory, curiously wrought everywhere with gold and silver and orichalcum; and all the other parts, the walls and pillars and floor, they coated with orichalcum. In the temple they placed statues of gold: there was the god himself standing in a chariot – the charioteer of six winged horses – and of such a size that he touched the roof of the building with his head; around him there were a hundred Nereids riding on dolphins, for such was thought to be the number of them by the men of those days. There were also in the interior of the temple other images which had been dedicated by private persons. And around the temple on the outside were placed statues of gold of all who had been numbered among the ten kings, both them and their wives, and there were many other great offerings of kings and private persons, coming both from the city itself and from the foreign cities over which they held sway. There was an altar too, which in size and workmanship corresponded to this magnificence, and the palaces, in like manner, answered to the greatness of the kingdom and the glory of the temple.

Here we are given another important detail. The ten kings (or their descendants) now rule not only over provinces of the island, but also over 'foreign cities' or colonies. The narrative goes on to detail the regions around the palace and the central island and the luxuries enjoyed by the kings. Of significance are the measurements supplied in Greek stadia. There would be no need for a fiction to include specific measures such as this. Taken at face value all the dimensions given are huge, but if we could find one trustworthy measurement then we could perhaps determine all the others by ratio.

Measurements are not a detail that we should trust in examining mythology. They are *not* a potential 'fossil'. The reasons for this should be somewhat obvious. There was no standard length for the Greek stadium, which varied between

157m and 209m among the city-states; in later Egypt, it was usually the latter. Commentators on Plato's other works also complain of his exaggerated measures. Clearly, if the details came from early Egyptian records then the dimensions were not originally recorded in Greek stadia. Perhaps Egyptologists can tell us what units were used in the earliest Egyptian times.[12] Either Sonchis must have supplied a translation of the units or Solon did this himself. The ancient Egyptians could not have measured the lost island themselves and the ultimate source must have been from ancient seafarers or the supposed inhabitants. Therefore, we see here more than one source of error; and it explains why later commentators would inflate what was clearly a large island to the dimensions of an entire continent 'larger than Libya and Asia put together'.

The narrative of Critias then moves on to describe the city around the palace and its various constructions.

...they had fountains, one of cold and another of hot water, in gracious plenty flowing; and they were wonderfully adapted for use by reason of the pleasantness and excellence of their waters. They constructed buildings about them and planted suitable trees; also they made cisterns, some open to the heaven, others roofed over, to be used in winter as warm baths; there were the kings' baths, and the baths of private persons, which were kept apart; and there were separate baths for women, and for horses and cattle, and to each of them they gave as much adornment as was suitable. Of the water which ran off they carried some to the grove of Poseidon, where were growing all manner of trees of wonderful height and beauty, owing to the excellence of the soil, while the remainder was conveyed by aqueducts along the bridges to the outer circles; and there were many temples built and dedicated to many gods; also gardens and places of exercise, some for men, and others for horses in both of the two islands formed by the zones; and in the centre of the larger of the two there was set apart a race-course of a stadium in width, and in length allowed to extend all round the island, for horses to race in. Also there were guard-houses at intervals for the main body of guards, whilst the more trusted of them were appointed to keep watch in the lesser zone, which was nearer the acropolis; while the most trusted of all had houses given them within the citadel, near the persons of the kings. The docks were full of triremes and naval stores, and all things were quite ready for use...

Leaving the palace and passing out across the three harbours, you came to a wall which began at the sea and went all round: this was everywhere distant fifty stadia from the largest zone or harbour, and enclosed the whole, the ends meeting at the mouth of the channel which led to the sea. The entire area was densely crowded with habitations; and the canal and the largest of the harbours were full of vessels and merchants coming from all parts...

The account then moves on to describe the surrounding plain and island, from which it is clear from Plato's words that he too doubted the dimensions that had been recorded by his source.

I have described the city and the environs of the ancient palace nearly in the words of Solon, and now I must endeavour to represent to you the nature and arrangement of the rest of the land. The whole country was said by him to be very lofty and precipitous on the side of the sea, but the country immediately about and surrounding the city was a level plain, itself surrounded by mountains which descended towards the sea; it was smooth and even, and of oblong shape, extending in one direction three thousand stadia, but across the centre island it was two thousand stadia. This part of the island looked towards the south, and was sheltered from the north. The surrounding mountains were celebrated for their number and size and beauty, far beyond any which still exist, having in them also many wealthy villages of country folk, and rivers, and lakes, and meadows supplying food enough for every animal, wild or tame, and much wood of various sorts, abundant for each and every kind of work.

I will now describe the plain, as it was fashioned by nature and by the labours of many generations of kings through long ages. It was naturally for the most part rectangular and oblong, and where falling out of the straight line had been made regular by the surrounding ditch. The depth, and width, and length of this ditch were incredible, and gave the impression that a work of such extent, in addition to so many others, could never have been artificial. Nevertheless I must say what I was told. It was excavated to the depth of a hundred feet, and its breadth was a stadium everywhere; it carried round the whole of the plain, and was ten thousand stadia in length. It received the streams which came down from the mountains, and winding round the plain and meeting at the city, was there let off into the sea. Farther inland, likewise, straight canals of a hundred feet in width were cut from it through the plain, and again let off into the ditch leading to the sea: these canals were at intervals of a hundred stadia, and by them they brought down the wood from the mountains to the city, and conveyed the fruits of the earth in ships, cutting transverse passages from one canal into another, and to the city. Twice in the year they gathered the fruits of the earth — in winter having the benefit of the rains of heaven, and in summer the water which the land supplied, when they introduced streams from the canals.

In such a description, we may see a construction project that rivals that of much later capitals such as Angkor Wat, or Aztec Mexico; but then no one would have believed those places either – until they were found. The description of the rectangular plain is a crucial detail. Do not focus on the actual measurements. We may see that its shape was a 3 by 2 rectangle, having the longer sides north to south and the shorter dimension east to west across the city itself. The city faced towards the south, or was open to the sea in that direction.

The social and military organisation of the island is then related for us in greater detail.

As to the population, each of the lots in the plain had to find a leader for the men who were fit for military service, and the size of a lot was a square of ten stadia each way, and the total number of all the lots was sixty thousand. And of the inhabitants of the mountains and of the rest of the country there was also a vast multitude, which was distributed among the lots and had leaders assigned to them according to their districts and villages. The leader was required to furnish for the war the sixth portion of a war-chariot, so as to make up a total of ten thousand chariots; also two horses and riders for them, and a pair of chariot-horses without a car, accompanied by a horseman who could fight on foot carrying a small shield, and having a charioteer who stood behind the man-at-arms to guide the two horses; also, he was bound to furnish two heavy-armed soldiers, two archers, two slingers, three stone-shooters and three javelin-men, who were light-armed, and four sailors to make up the complement of twelve hundred ships. Such was the military order of the royal city – the order of the other nine governments varied, and it would be wearisome to recount their several differences.

We may see here another significant 'fossil' that aids our enquiry. The rectangular plain was divided-up not only by canals and irrigations, but also into square fields for different ownership, with other settlements in the highlands around. The military included naval forces, as we might expect for an island. We may pass over here the description of the laws of Poseidon, which expand on the previous statements. We then find one of the most useful of the fossilised details in the myth:

> …Now the order of precedence among them and their mutual relations were regulated by the commands of Poseidon which the law had handed down. These were inscribed by the first kings on a pillar of orichalcum, which was situated in the middle of the island, at the temple of Poseidon, whither the kings were gathered together every fifth and every sixth year alternately, thus giving equal honour to the odd and to the even number. And when they were gathered together they consulted about their common interests, and inquired if any one had transgressed in anything, and passed judgement…

This statement: that the kings of the various provinces gathered every alternate *fifth* and *sixth* year is concise; and it tells us that the Atlanteans had a calendar. This calendar has been examined in depth in the present author's earlier works and need not be revisited here in such detail.[13] Briefly, there are only a few ways that lunar months and solar years can be arranged to make a repeating lunisolar calendar cycle; this formula of 11-years divided into 5- and 6-year periods is one such method. We may see that, quite unlike the measurements, such a simple numerate detail can pass through translation and retelling without loss of accuracy. Plato did not need to devise a real lunisolar calendar just to provide background for a fiction.

We are next given a description of bull sacrifice, a rite that has parallels elsewhere in the ancient world and survives in various forms to the present day in the bull fighting of Spain and France:

> There were bulls who had the range of the temple of Poseidon; and the ten kings, being left alone in the temple, after they had offered prayers to the god that they might capture the victim which was acceptable to him, hunted the bulls, without weapons, but with staves and nooses; and the bull which they caught they led up to the pillar and cut its throat over the top of it so that the blood fell upon the sacred inscription. Now on the pillar, besides the laws, there was inscribed an oath invoking mighty curses on the disobedient. When therefore, after slaying the bull in the accustomed manner, they proceeded to burn its limbs, they filled a bowl of wine and cast in a clot of blood for each of them; the rest of the victim they put in the fire, after having purified the column all round. Then they drew from the bowl in golden cups, and pouring a libation on the fire, they swore that they would judge according to the laws on the pillar, and would punish him who in any point had transgressed them...

The remainder of this account describes the ceremony by which the various kings would promise to come to each other's aid and not make war on each other; and to curse those who might transgress the laws set down by Poseidon; feasting all day and burning and dedicating even their robes to the sacred rites.

> Such was the vast power which the god settled in the lost island of Atlantis; and this he afterwards directed against our land...

The Critias ends shortly after this point and was clearly unfinished. The promised dialogue of Hermocrates is not given and the account of the war between ancient Athens and the kings of Atlantis; and the social order of the earliest Athenians was never written – although we may assume that the unfinished fragment we have was only to be the introduction to this: the primary interest of its author.

For other independent verification of these ancient events we should start first with the earliest commentators and writers on related subjects; in particular those who may have had access to other lost books and sources. Diodorus Siculus, as we have seen, specifically declined to comment on Plato, although he still gives us much that is useful. In particular, he supplies a rationalisation of the Greek pantheon of gods. Most of what we are told of them in poetry and in the encyclopaedia of Apollodorus (first century BC) would fit well with the disparaging view of Sonchis the Egyptian regarding the Greeks' jumbled knowledge of their own past: 'Oh Solon', he says, 'you Hellenes are never anything but children and there is not an old man among you…' Yet many other sources state the equivalence of the myths and traditions between Greece, Egypt, Rome and other nations. After his account of the Amazons, Diodorus goes on to say:

> But since we have made mention of the Atlantians, we believe that it will not be inappropriate in this place to recount what their myths relate about the genesis of the gods, in view of the fact that it does not differ greatly from the myths of the Greeks. Now the Atlantians, dwelling as they do in the regions on the edge of the ocean and inhabiting a fertile territory, are reputed far to exceed their neighbours in reverence towards the gods…And their account, they maintain, is in agreement with that of the most renowned of Greek poets [there follows a quote from Homer's Iliad].[14]

However, this thread need not be pulled further in preference to an examination of the earliest geography and ethnology of the region.

Other Writers on Libya and the Atlantians

As we have seen, Diodorus was equally disparaging of Herodotus, who wrote some four hundred years before him; he too visited Egypt about a hundred years after Solon and he too spoke to the priests there. He gives us an account of the tribes of North Africa as they then existed and of the tribes around the river Triton, he records

They sacrifice to the sun and moon, the worship of which is common to all the Libyans, though those who live round Lake Tritonis sacrifice chiefly to Athene and after her, to Triton and Poseidon. It is evident, I think, that the Greeks took the 'aegis' with which they adorn the statues of Athene from the dress of Libyan women...[15]

Earlier he has described the tribes further to the west, among them are the *Atlantes*, named after the Atlas-mountain but he is otherwise imprecise about the geography. He says that they eat no meat and they never dream!

A little before that he describes the worship of Athene among he Machlyes people who lived along the River Triton:

The Machlyes...hold an annual festival in honour of Athene, at which the girls divide themselves into two groups and fight each other with stones and sticks; they say this rite had come down to them from time immemorial, and by its performance they pay honour to their native deity – which is the same as our God Athene. If any girl, during the course of the battle, is fatally injured and dies, they say it is proof that she is no maiden...there is a belief among these people that Athene is the daughter of Poseidon.[16]

He also tells us:

...the women of the tribe are common property; there are no married couples living together, and intercourse is casual – like that of animals.

Now Herodotus is here showing a typical male-biased viewpoint when confronted with a matriarchal culture, where it is really the women who make the decisions!

This equivalence of Neit and Athene is of course a unifying aspect between the myths of Diodorus and Plato previously related above. Another name of Athene was *Tritogeneia* 'Triton-born' because she sprang from the earth and was found beside the River Triton; the myth then continues:

> ...this goddess, choosing to spend all her days in maidenhood,
> excelled in virtue and...she cultivated also the arts of war

It may be that we see here another of those instances where
legendary and mythical heroes (or heroines) of similar name
and achievements have become amalgamated into a single
character upon the degradation of the history.

One may compare all this with the description of the
warlike Amazon women; and the relationship of Athene to
Poseidon confirms that Solon's story came via Egyptian rather
than Greek sources of mythology. Herodotus supplies a further
insight; the names of the Greek gods, he says, all came from
Egypt:

> ...with the exception...of Poseidon and...I have the authority
> of the Egyptians themselves for this. I think the gods of whom
> they [the Egyptians] profess no knowledge were named by the
> Pelasgians – with the exception of Poseidon, of whom they
> learned from the Libyans; for the Libyans are the only people
> who have always known Poseidon's name, and always
> worshipped him...[17]

Here, we see another example of a 'fossil' in the myths,
checked against an independent source.

Proclus Lycaeus of Byzantium (AD 412-485) was one of
the so-called Neo-Platonist philosophers who built on Plato's
work. He composed a long commentary on the *Timaeus* and
supplies fragments of earlier writers, in particular, Crantor,
Porphyry of Tyre and a historian named Marcellus. Coming
from historians much closer to the events than ourselves, such
sources deserve our respect. His commentary focuses upon the
catastrophes and the astronomy rather than so much
mythology.

Marcellus is otherwise unknown, but he may have written
earlier than or quite independently of Plato a history of Africa
called *Aethiopica*. Certainly, we should expect such a work to
include Libya and Egypt. On Atlantis, he is thus quoted:

> That such and so great an island existed, is evinced by those
> who have composed histories of things relative to the external

sea. For they relate that in their times there were seven islands in the Atlantic sea, sacred to Proserpine: and besides these three others of immense magnitude; one of which was sacred to Pluto, another to Ammon, and another which is in the middle of these, and is of a thousand stadia, to Neptune. And besides this, that the inhabitants of this last island preserved the memory of the prodigious magnitude of the Atlantic Island, as related by their ancestors; and of its governing for many periods all the islands in the Atlantic sea. And such is the relation of Marcellus in his Aethiopic history.[18]

Figure 1.3 The Megalithic Culture Region of the Middle-Neolithic, showing the principal areas where megalithic monuments are found.

The translation of the names of the gods into their Roman equivalents would argue against those who say Marcellus wrote before Plato, but perhaps Proclus translated the names. Pluto is the Greek Hades, god of the underworld; Ammon is Zeus, Proserpine for Greek Persephone; and Neptune is Poseidon, god of the Sea.

However, that the source is independent of Plato and Solon is clearly stated. It comes from other historians who wrote specifically about the Atlantic coast. It gives us a size for the central island in stadia, which should be free of any mistranslation of units found in Plato. If the central island was a thousand stadia (long?) then it would make it about the size of Cornwall, or about twice the size of the Isle of Man. The two islands around it were much larger (we are told) but certainly not 'continents'; and for the seven other islands no dimensions are offered. It will be interesting to find all these islands along the coast of Europe – for these are not questionable lands sunk in a disputed myth; these were extant real islands contemporary with the cited historians. One of these islands preserved a myth, which echoed that found in Plato's narratives.

As so often with myths and ancient history, we may despair that such valuable historical sources are lost to us. Similarly the work of Crantor is unknown but for a few citations. As one of the earliest commentators, his words are often misquoted by more-recent enthusiasts. In the commentary of Proclus however, his actual recorded words are somewhat more cautious.

> Crantor adds that this is testified by the prophets of the Egyptians, who assert that in these particulars [which are narrated by Plato] are written on pillars which are still preserved.

Now it seems hard to believe that all of the information we have examined, which allegedly comes via Solon's notes, should be originally condensed to hieroglyphs on temple walls, including, for example, the trivia of which fruits the god-kings used to eat. Perhaps, therefore the hieroglyphs were used to aid memory of an oral history, passed down the generations. We shall examine in the next chapter what Crantor might actually have seen.

Of these 'pillars' you may often see modern references in internet sources, etc, that these were the so-called 'Syriadic Columns'. In fact, these latter are mentioned by the Jewish historian Josephus rather than by Crantor and we cannot be sure that they were the same monuments.[19] The idea that records from before the time of the Flood were preserved on such columns also comes from Manetho (or someone claiming to be him) as recorded by Syncellus. However, all this is very tenuous. Other than confirmation that the Egyptians recorded ancient history on stone monuments, this supplies little extra detail.

Manetho is another independent source that we may consult for verification. For Egyptologists his king list is the very basis of the Egyptian chronology as he grouped the reigns into the thirty 'dynasties' that are not given in other surviving king lists. The king lists are perhaps an aside from the present analysis, save for their similar preservation on temple walls and columns. However, the chronology can be explored further in the next chapter.

To close then, one may summarise the evidence and the coincidences that are found in the sources. The story of Atlantis, its colonial empire and its destruction was claimed to be a genuine history brought to Greece by Solon from Egypt; its detail came via written notes and ultimately from an Egyptian Priest of Saïs, who gave translation. Other historians knew of similar stories about a great empire that had once existed on an island in the Atlantic Ocean.

Saïs was a real city, on the Canopic branch of the Nile: the modern Sa El Hagar. In this city was the shrine of the goddess Neit, who was equated with Greek Athene. We also see in these sources that Plato's description of their deities concurs better with Egyptian-Libyan and North African versions rather than the Greek. That ancient Egyptian history was recorded in sacred records (at Saïs or other temples) is confirmed by comparison with Diodorus Siculus and Herodotus. The religion of Neit is associated with the invading Amazons whose homeland was similarly lost to the sea. Diodorus independently describes a people called Atlantians living in North Africa and that the gods were believed to have been born among them. Herodotus also confirms tribes around 450 BC who practiced similar religion and matriarchal customs; and who may have been the descendants of the ancient peoples described in the myths.

Despite the clarity and detail in the description of an ancient society that we have inherited via Plato, it should not be regarded as 100% accurate in every detail. If accepted at face value it is the interpretation of a Greek author, from a set of notes written in Greek and derived from an Egyptian written source; and which describes a place distant in time and space even when it was recorded. Even so, when Solon received the information it was a written history, not a vague myth or legend. It is a legend to us only because we do not have the original source against which to check it.

Such then, are the 'elements' that will be used in our experiment. We can now move on to see what other substance we can add to this base mixture of myths and legends.

Notes and References

[1] In my earlier works *The Atlantis Researches* or *Atlantis of the West,* the passages were quoted in full from Benjamin Jowett's excellent translation of 1892 so there is no need to do so again here.

[2] Plutarch, Life of Solon, 26.

[3] May we therefore presume that Plato had actually seen Solon's notes?

[4] These events and advances have all occurred since my earlier books in the 1990's, in which it was necessary to devote an entire chapter to justification of this catastrophist astronomy; such groundwork should no longer be needed.

[5] Diodorus Siculus I. 27, 28

[6] Diodorus Siculus I. 69, 6-70; Oldfather's translation

[7] These women warriors should not be confused with the Amazons of Scythia mentioned elsewhere in Greek mythology.

[8] This is another example of a detail that is not needed to tell a fictional story.

[9] The name 'Libya' in classical sources is not restricted to the modern country; it is used vaguely and could describe any part of the North African coastal region; with the general name 'Ethiopians' similarly used of any black African races further south.

[10] This is the same city from which Diodorus's source: Dionysius 'Skytobrachion' came. Perhaps it is not surprising that he should have researched the history of his own island.

[11] Diodorus Siculus Book III.54.55; Oldfather's translation

[12] In later Egypt the royal cubit was about 28 fingers long; the rod of chord: 100 cubits; and the 'river measure': 20,000 cubits – but usage in early dynasties cannot be certain.

[13] In an appendix to *The Atlantis Researches* or *Atlantis of the West*; and in Chapter 5 of my *Under Ancient Skies* and other detailed papers, where a full reconstruction of the lunisolar calendar is given. It is this coincidence of astronomy more than any other that should convince us that the Atlantis story is authentic ancient history.

[14] Diodorus Siculus II. 56

[15] Herodotus, IV, 188

[16] Herodotus, IV, 181

[17] Herodotus, II, 50

[18] This translation by Thomas Taylor, Plato the Timaeus and Critias, Bollingen Series 3, reprinted by Pantheon Books, Washington (1944); he also gives the original Greek text. A similar translation by Thomas Taylor (1820) is given in my earlier books.

[19] Josephus: *Antiquities of the Jews* Book I. 2, 158.

2

History Forgotten –
When was the Atlantis Era?

Anyone who studies the German language will know that the Germans always organise their sentences strictly in the order of *time-manner-place*; and so a similar rule may be followed here in the analysis of the ancient catastrophe. *Vor Tausenden von Jahren versank Atlantis in den Atlantischen Ozean*: 'Thousands of years ago Atlantis sank into the Atlantic Ocean'. So let us begin with the *when*.

Firstly to be clear that, be it true or otherwise, the origin of the story is Egyptian and not Greek. Plato says that Solon brought back the detailed account after visiting Egypt and talking directly with the priests of Saïs. Classical scholarship usually ignores this internal evidence and seeks to dismiss it as a purely Greek fiction devised by Plato himself to contrast its dynasty of kings with the democracy of Athens.

The date that has attracted all the attention has always been the statement, of the Egyptian priest Sonchis of Saïs: that the events occurred 8,000 years before his own time. This fabulous chronology has been a major cause of the disbelief even among the ancient commentators. Much less attention has focused upon his other statement: that it was contemporary with the earliest period of the Egyptian state.

> She [Neit/Athene] founded your city [Athens] a thousand years before ours [Saïs]...of which the constitution is recorded in our sacred registers to be 8,000 years old.

So we may properly deduce that the real date of these events was contemporary with the founding of Saïs and the civilization of Lower Egypt (the Nile Delta).

Of course when modern Egyptology began no-one really knew how old Egypt was. There were competing long and short chronologies – but most scholars will now agree on the short chronology and a date around five-thousand years ago for the unification of the 'Two Lands': Lower Egypt – the Delta region and Upper Egypt – the civilization of the Nile valley. This puts the establishment of the First Dynasty and a King Menes or Horus Aha at around 3100 BC, with a predynastic or proto-dynastic period of perhaps two hundred years before that. So this would be the proper era to seek evidence of a geological catastrophe and to look for similarities with other myths and cultures around the world.

The theory that has come to pervade most academic discussion and the comfortable non-controversial output of books and television programmes has been to equate Atlantis with the known destruction of Crete and Akrotiri by the nearby eruption of the Thera (Santorini) volcano. This theory dates from the discovery of the ruins of the buried city on the island of Santorini in the 1960s by Professor Marinatos, who immediately saw the parallels with Plato's description. The academic world seized upon this discovery as a safe rationalization for Plato's myth. The theory is put about as if proven fact – yet even the application of a moment's common sense will cast doubt on this explanation.

The latest archaeological dates for the civilisation on the island of Crete will place it between 1950 BC and 1450 BC; and the volcanic destruction of Thera around 1645 BC. This explanation has to rely on the dismissal of the story as Plato's fiction. Whatever date is adopted for Thera it is certainly more than a thousand years later than 'the beginnings of the Egyptian state'; indeed the Egyptians well knew of their Cretan neighbours and traded with them during the Pyramid age. Other writers would like to compare the Thera eruption with the events of the Biblical Exodus, which is conventionally ascribed to the New Kingdom era during the reign of Ramesses the Great.

Similarly, the belief that Atlantis and the catastrophe that submerged it can safely be hidden away 10,000 years ago at the end of the ice age has long satisfied the popular authors and their publishers. So in rejecting the statement of Sonchis in this single regard, and yet relying upon his other descriptions, it is perhaps best to again quote the opinion of Diodorus Siculus, which is free from any modern bias:

> Some of them [the Egyptian priests] give the story that at first gods and heroes ruled Egypt for a little less than eighteen thousand years, the last of the gods to rule being Horus, the son of Isis; and mortals have been kings of their country, they say, for a little less than five thousand years...[1]

And also;

> The priests of the Egyptians, reckoning the time from Helius [the sun's reign] to the crossing of Alexander into Asia, say that it was in round numbers twenty-three thousand years.[2]

Diodorus was sceptical of these very long reigns and goes on to say that the reckoning of time was formerly counted in lunar months, or perhaps in seasons, before the period of the year was properly understood; and indeed we find this preference for lunar calendars in other cultures worldwide. For the same reasons, we may reject the ancient priests' calculation of era as it would include part of these divine reigns of gods, demi-gods and spirits of the dead.

Far more useful for dating purposes is the simple statement of Sonchis that history was written down and preserved in their temples from the earliest period of the Egyptian state. Egyptologists will assert to within an accuracy of decades that the earliest known written hieroglyphs date from the First Dynasty around 3100 BC; with the events described in the myths being therefore assigned to the late fourth millennium BC. So what other evidence may we cite to accurately date the founding of Lower Egypt, the temple of Saïs, and the religion of the goddess Neit?

Manetho and the Egyptian King Lists

Following the campaigns of Alexander the Great against the Persian Empire and his conquest of their Egyptian province in 332 BC, Egypt fell under the influence of the Greek dynasty of the Ptolomies, which would persist right through to Cleopatra. While Egyptian culture continued little disturbed under its new rulers, the Greeks could not read its hieroglyphs and cursive scripts. It was for this reason that around 280 BC a native Egyptian whose name is recorded as *Manetho* was commissioned to write an explanation of the king lists in Greek known as the *Aegyptiaca*;

Some scholars would like to translate Manetho's Egyptian name as 'Beloved of Thoth', or perhaps 'Beloved of Neith'. His original text is lost, but it can be reconstructed from the fragments recorded by later copyists – all of which differ in the names and reign lengths; and even in the lengths of the dynasties. Some commentators even doubt that Manetho ever existed and would prefer to call him 'Pseudo-Manetho', being a creation of a later copyist; but to recount here the preservation of the king lists through these later writers would divert from the central task. Suffice to say that although Manetho may have been a native of Sebennytos in the Delta, we do not find anything that would suggest he consulted local sources, or indeed anything that might confirm the history given to Solon. Egyptologists however do prefer to see inspiration from Lower Egypt in some of Manetho's later dynasties.

Manetho was also cited by Josephus in his Jewish history. However, any link between his so-called Syriadic Columns, or 'columns of Seth' (on which antediluvian records were supposedly preserved) with those 'pillars' that may have held the Saite sacred texts, would introduce only further confusion.

The long chronology of divine rulers that evidently influenced Sonchis of Saïs is confirmed in the introduction to the Aegyptiaca. The god Thoth (Greek Hermes) was credited with the invention of writing and a translation is implied from 'sacred characters' into Hieroglyphs by his divine-son Hermes Trismegistus at some remote period. These formed the basis of the 'sacred books', which in Ptolemaic times Manetho would translate. However, other than a coincidence of the name it is unsafe to equate these books with the sacred texts mentioned in Plato's narrative or with those that Diodorus used to check his history.

Manetho's king lists begin with the dynasty of gods, the last of whom was Horus. The gods ruled Egypt for 13,900 years; followed by a dynasty of demi-gods (deified mortal kings) for 5,212 years; and then by 'sprits of the dead' for 5,813 years: total 24,925 years. Other than the gods, no names are given. These were followed, says the text, by five native Egyptian tribes who ruled the Nile valley before the dynasties of kings.

Together with the reconstructed Manetho dynasties, modern Egyptologists have been able to put together a chronology for Egypt by comparing it with other available king lists. One source is the fragmentary Royal Annals, a Fifth Dynasty Stele believed to come from Memphis, of which the largest fragment is the Palermo Stone, Most Egyptologists concur that the early part of Manetho's list agrees best with the reigns recorded there, while later reigns are closer to the Turin Canon. Other sources: the Karnak Tablet, the Abydos Tablet, the Saqqara Tablet and the fragmentary Turin Papyrus confirm kings back to the First Dynasty and begin with a king named *Menes* who was the first to rule over both Lower Egypt (the Nile Delta) and Upper Egypt. However, the older Palermo stone gives different names and calls the first king only by the name *Horus-Aha*. It also offers some names of predynastic kings that are not found elsewhere.

Herodotus also tells us of Menes, or *Min* whom he describes as the first king of Egypt; and who ruled at a time when the Nile Delta first rose out of the Sea. He could observe himself that the Nile Delta extended into the sea beyond the coastline on either side. For the Egyptians, 'Libya' began west of the Nile with 'Arabia' to the east. Herodotus believed his hosts when they told him that the Delta had formed only in historical times:

They...went on to tell me that the first man to rule Egypt was Min, in whose time the whole country except the district around Thebes, was marsh, none of the land below Lake Moeris...then showing above the water.[3]

Now interesting though this is it does not confirm that Menes was the unifier of the Two Lands. He could equally have been just the first king of Upper Egypt and therefore a predynastic ruler. After all, if the Nile Delta had not yet formed in the time of Menes then what was there for him to unify? Logically, Lower Egypt did not then exist! Egyptologists conveniently ignore this detail: it's only Herodotus and his mythology again; and who was he but a historian closer to the events than we are and who spoke to people who kept historical records of their own past?

Archaeology can confirm that human occupation of the Delta began about 5,000 years ago – around the time that Menes was combating river floods. Records surviving from Dynasty I show a decline in the height of the annual inundation during this dynasty. Sites of occupation that may be earlier are found only on the so-called 'koms' or mounds that today remain above the flood plain of the western Delta.[4] In the earliest times these may have been just islands in a marshy estuary of the Nile.

Horus-Aha and Menes
Egyptologists will usually accede that Menes is the same person as the King Min described by Herodotus. As so often archaeologists make use of myths and legends when they confirm their opinions yet dismiss them entirely when they do not. For example, it is depressing when Walter Emery, an eminent authority whom one must cite, dismisses the legends cited by Diodorus and Manetho:

> Traditional stories concerning Egypt's first pharaoh are recounted by Diodorus but are hardly credible and of little value. According to this classical writer, the king, while hunting in the Fayum, was treacherously attacked by his dogs and only escaped by jumping into Lake Moeris, where he was carried to the opposite shore by a crocodile.[5]

It would be refreshing to hope that the attitude of archaeologists towards mythological evidence has improved since the 1960s.

Egyptologists however, can find no trace of a king named Menes on any monuments or artefacts that can be dated to the First Dynasty. The standard chronology depends crucially upon counting back the agreed reign lengths in the king lists (now variously verified by radiocarbon) from the earliest 'fixed point': a reference to the Egyptian Sothic calendar in the reign of Senusret III. This calculation assumes that Menes must be the same person as Horus-Aha. A margin of error around 150 years attaches to the chronology, due to the uncertainty of reign lengths, overlapping co-reigns and differing opinions about the accuracy of radiocarbon dates. However, the long-standing consensus that the unified Egyptian state dates from the end of the fourth millennium BC is not challenged. A summary from the best available opinion (as at 2013) would be:

Late Predynastic Period	c.3500-3100 BC
Early Dynastic Period	c.3100-2686 BC
'Scorpion'	
Narmer	
(First Dynasty)	c.3100-2890 BC
Hor-Aha	
Djer	
Wadj (Djet)	
Den	
Anendjib	
Semerkhet	
Qa'a	

These are the so-called Horus-names of these kings, but Egyptian kings had many titles, so the fact that we don't see the name Menes is unremarkable.

At the end of the nineteenth century when archaeologists excavated the First Dynasty *mastaba* tombs, at Abydos in Upper Egypt; and at Saqqara at the head of the Delta they found that all had been robbed long ago. The earliest excavators, who were scathingly criticised by Sir Flinders Petrie, must have ignored much more as he recovered many artefacts from their rubble. He described eleven tombs including those of *Narmer*, *Horus-Aha* and later kings of the dynasty; and also that of a queen whom he called *Merneit*, or by later Egyptologists *Merneith* or *Meryet-nit*. We may find the name of the goddess Neit associated with other rulers of the archaic dynasties, thus proving her ascendency at this early period. It was Petrie and later eminent Egyptologists such as Reisner and Emery, to whom modern researchers all refer back, who would build upon the linkage of Menes with Horus-Aha and Narmer. It is important to appreciate that this is eminent opinion and not proof.

Narmer or Mer-i-Nar?

James Quibell who assisted Petrie at the predynastic site at Hierakonpolis can claim discovery of the Narmer Palette, although precisely where was not recorded. The decoration is considered to be of an archaic style but it may actually have been buried much later. The rounded-triangular slab of slate appears to show on one side a king wearing the 'red crown' of Buto and Saïs, which would later come to represent the king of Lower Egypt. The reverse shows the king (or perhaps a different king?) wearing the white crown of Upper Egypt and smiting a defeated enemy with a mace. At the top of the palette is an early form of hieroglyphics that the specialists decipher as 'Narmer' – the name of the king. However, since hieroglyphs give only the consonants, and could sometimes be read in either direction, they can actually give us no sense of how it was pronounced; all we actually have is *n'r* and *mr*. If read the opposite way it could equally be *mr* and *n'r*.[6] The palette is usually regarded as a commemoration of the victory by a southern king over the northern king to unite the Two Lands. Therefore we have the equation: Narmer = Horus-Aha = Menes. Again it boils down to eminent opinion that lies deep in all the text books.

A decorated mace-head discovered at Hierakonpolis has also attracted controversy among Egyptologists, for it appears to show Narmer wearing the crown of Lower Egypt greeting an important person whom Egyptologist Walter Emery interpreted as a queen.[7] He would see in this Narmer who as: 'conqueror of the North attempted to legitimize his position by taking the Northern princess as his consort'. Hence it has come to be known as the Narmer 'wedding' Mace head. Emery believed she was the 'Neit-hotep' (Neit-is-satisfied) known from an ivory plaque discovered at Helwan. Of course, it is acceptable for an eminent archaeologist to speculate in this way from such scant evidence.

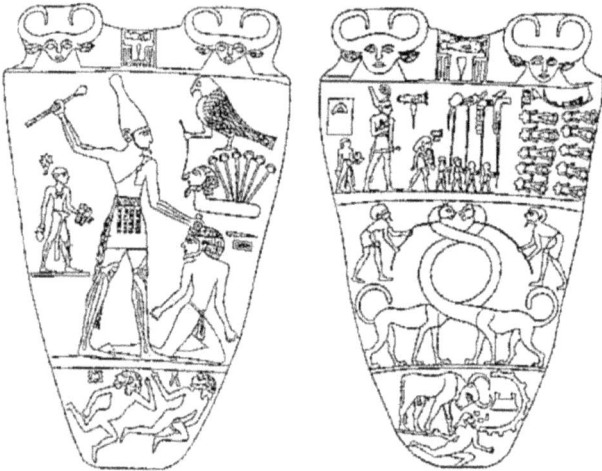

Figure 2.1 Drawing of the Narmer Palette showing serekh on each side and repeated above the procession of the king

Other inscriptions on artefacts from Hierakonpolis supply more details for the specialist to interpret. A mace head of white limestone shows a similar king wearing the white crown of Upper Egypt, who is digging a canal. A hieroglyph of a scorpion precedes the image of the king and because he wears only the white crown he has been interpreted as a pre-dynastic king named Scorpion. Hence this early period is sometimes termed Proto-dynastic, or Dynasty 0.

From the rubble at Abydos, Petrie recovered two ivory labels with identical inscriptions. Four rows of early hieroglyphs are interpreted to show the name 'Aha' at top left, two sacred boats and then a recognizable symbol that (in later inscriptions) would depict the temple shrine of the goddess Neit. Two further rows show a story that cannot really be understood and the fourth row is a continuous series of unintelligible hieroglyphs, which Petrie would describe as the earliest example of its kind.[8] It is this precocious use of hieroglyphs which convinces Egyptologists that the invention of writing predates the First Dynasty.

At this early period the king's names were not yet placed in a cartouche. Usually, a square serekh preceded by the Horus-falcon symbol contains the Horus name of the king. More recent discoveries in the Abydos tomb of King Den give us another early king-list. A seal impression lists the six names of *Narmer, Aha, Djer, Djet, Den*, and a queen *Merneith*. In the tomb of Qa'a a similar seal shows eight kings: *Narmer, Aha, Djer, Wadj, Den, Anendjib, Semerkhet*, and *Qa'a*, this time with the queen omitted. In later times it would become the norm for female rulers, such as Hatshepsut and Nefertiti, to be dropped from king lists. The specialist interpretation is that the archaic kings would take a Horus-name while their queens would take a name associated with the goddess Neit: (Merneit

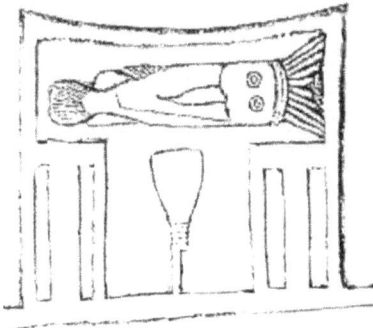

Figure 2.2 Serekh of 'Narmer' on the Narmer Palette. The symbols are interpreted as *n'r* (catfish) and *mr* (chisel). Equally they could be read (properly or mistakenly) in the reverse direction.

means something like Beloved-of-Neit). It is therefore assumed that the southern king is taking a northern princess as a bride in order to legitimize his rule in the north, but if so then who are the northern kings? The male-centric viewpoint ignores the possibility that these were matriarchal queens, inheriting the northern throne in their own right, along the female line of descent.[9] This would conform to the free social order that ancient historians tell us prevailed among the Libyan tribes; it would also explain the custom of brother-sister marriage as a device to retain succession within the ruling dynasty.

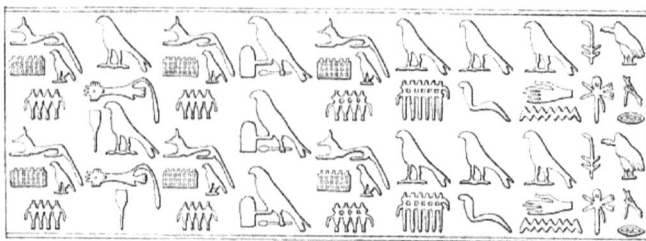

Figure 2.3 Cylinder seal impression from the tomb of Horus Den
This shows a First Dynasty king list; from left to right are serekh of
Narmer, Aha, Djer, Djet, Den and Merneith. Note again the falcon implying
the Horus-name of the king, with at left the catfish and chisel symbol of
'Narmer'; and at right the crossed-arrows symbol of Neit, implying a queen.
(From W. Kaiser in M.D.A.I.K. 43, 1986, p. 115-119).

In addition we may note that numerous *serekh* bearing the
name of Narmer have been found by archaeologists at nine
different sites in Canaan, indicating an Egyptian presence in the
Levant at this early date.[10] Some authorities even claim that
Egyptian influence in Canaan peaked at this era. In Egyptian
sources anything east of the Nile was absorbed in the loose
term 'Arabia' and so is difficult to pin down. This presence has
contributed to Egyptologists' convictions that Narmer was the
unifying king from the south who conquered the north by force
and went on to build an empire.

Neith: a goddess by any other name...

To explore a little further about the goddess *Neit*; that form is
preferred here because it was used by Jowett in his translations
of Plato as quoted previously, but you will also see *Neith*, *Nit* or
Net in other contexts. Plato's narrative itself equates her with
the Greek goddess Athene; and hers is the only Egyptian name
that Solon preserved when he translated all the others.

Herodotus visited the temple of Neit in 454 BC and gives us
a description of its interior as it was about 140 years after
Solon's visit. He tells us how king Amasis had built a
remarkable new gateway and describes the tombs of the Saite
kings in a precinct close to the shrine of the goddess; and also
the shrine of Osiris – whose name he will not speak due to his

associations with Greek Dionysus and his unseemly ceremonies. Of the architecture we may note:

> ...the tomb of Amasis too...is in the temple court, a great cloistered building of stone, decorated with pillars carved in imitation of palm-trees. Here too is the tomb [of the unmentionable one]...it stands behind the shrine and occupies the whole length of the wall. Great stone obelisks stand in the enclosure.

We find no direct confirmation here, just a century after Solon, of any sacred inscriptions or histories written on the obelisks, but we may be sure that the walls and columns would be decorated with appropriate religious texts. Frustratingly, Herodotus tells us later, when visiting the temple of Hephaestus at Memphis, that the priests there read to him from a list of three-hundred and thirty monarchs. He also tells us that King Min built a dam and diverted the river where he then founded the city of Memphis. It is more likely that this was the source of the king lists that Manetho would later translate to Greek.

Fig. 2.4 Neit

For archaeologists, almost nothing of ancient Saïs survives to be excavated at Sa El Hagar; it being long since destroyed by farming and the removal of so much sediment as raw material for mud-bricks, or simply washed away by the Nile floods. We may hope that one day some archaeology may be discovered. The best we can assume is that in predynastic times it was an island, which was occupied early during the emergence of the Delta.

As to who built the temple and precisely when it was founded, the association of the early kings with Neith-queens has led the Egyptologists to assume that it was founded during the 'unification'; in which Horus-Aha or Menes subjugated the northern kingdom by force of arms – but we don't actually know any of this! It cannot be older than the

Delta itself. Plato's narrative only tells us that the goddess herself founded the temple. That she was originally a Libyan goddess worshipped all along the Libyan coast to the west is indicated by the numerous myths associating her with Athene.

Such as is known from Egypt itself of the religion of Neit comes from her symbols in later Dynastic times. The ascendancies of the various Egyptian deities would rise and fall many times over three thousand years, but specialists recognise the name and symbols of Neit again and again. The hieroglyph of her name shows two bows bound together; or a shield with crossed bows as seen in the early dynastic period. It is this warlike association that underscores the later Greek belief of a common origin with their own virgin warrior-goddess Athene. By the time of the Old Kingdom pyramid age she had already become absorbed into the convoluted pantheon of Egyptian gods and is often indistinguishable from Isis. An examination of all these associations only adds the mystical confusion that comes with religious beliefs and we lose any hope of getting at real history.

The Saite Dynasty – Native Egyptian Culture Reasserted

The era when Solon visited Egypt was around 590 BC probably within the reign of Psammetichus II. Around 664 BC Psammetichus I had defeated the invading Assyrians and Nubians and established the Twenty-sixth Dynasty capital at Saïs. This dynasty was to be the final flourish of independent Egyptian culture before it fell successively under Persian, Greek and Roman dominion. During this Saite Dynasty the native traditions of the First Dynasty were revived and the temple at Saïs was greatly embellished by the successive rulers who built their tombs there.

Herodotus tells us that a profitable commerce developed between Greece and Egypt under the Saite rulers and he describes the many privileges offered to Greek traders and travellers. King Amasis spent this increasing wealth on lavish restoration of the temple – one room, we are told, was hollowed from a single block of stone brought from the upper Nile – and he added numerous Sphinx statues. Comparable works were undertaken at other temples along the Nile

Such a wealthy country made a tempting target for the invading Cambyses in 525 BC. Citing insults by the Egyptians the Persian king advanced – to be faced by king Psammetichus III. For his father Amasis (Ahmose II) had recently died of natural causes after a forty-four year reign of unparalleled prosperity and was given lavish burial in the tomb he had prepared for himself in the temple of Neit. Herodotus describes a hard battle, with equal casualties on both sides, but it was the Egyptians who broke and fled to Memphis. Psammetichus, we are told, drank bull's blood when further resistance was clearly hopeless. Cambyses then went to the temple at Saïs and treated the body of Amasis to great insult before having it burned – the worst thing that could happen to an Egyptian king, for it denied him the afterlife.

Following desecration of all the Egyptian temples by the Persian invaders, Cambyses allowed the temple of Neit to be restored and its rites continued to be performed throughout the Persian hegemony and under the later Ptolomies.[11] It was only during these late dynasties from the twenty-sixth onward, with the Delta again dominant over the south, that Neit regained the prominence of two-and-a-half thousand years earlier.

The date of 525 BC has significance for another reason; the invasion of Cambyses is the earliest event that can be tied with certainty to the modern Gregorian calendar. Before that date history increasingly becomes guesswork and depends upon counting back the reigns in the king lists. Before radiocarbon, cross-dating of artefacts back to the Egyptian chronology was the only way that archaeologists had to date ancient cultures.

The Elysian Fields: a Paradise in the West

A student of Plato's school named Crantor (c.335-275 BC) visited Egypt and attempted to verify the sources of Solon and Plato. He also was told of (or perhaps himself saw) inscriptions on pillars that he recognised to be the same as described by Solon, but we cannot be sure that the pillars he mentions were at Saïs. His words are quoted in a commentary on Plato by the later Neo-Platonist philosopher Proclus: 'this is testified by the prophets of the Egyptians, who assert that in these particulars...are written on pillars which are still preserved.'

Egyptologists have long known of the ritual spells or prayers that were written on papyrus scrolls and buried along with kings and other prominent persons, known variously as *The Book of the Dead*. These were reproduced as part of the burial rites of wealthy citizens but one doubts whether the later artists and scribes really understood what they were all about. Egyptologists soon recognised that some of these utterances were similar to those inscribed on the walls of the Fifth and Sixth Dynasty pyramids at Saqqara (2494-2184 BC) but unlike later versions of the spells, the Pyramid Texts were not illustrated.

The Egyptian name for these religious texts was 'The Chapters of Coming Forth by Day'. These underwent a series of revisions or recensions during the long history of Egypt; Egyptologists recognise the utterances in the Pyramid texts as the oldest version and have termed this the *Heliopolitan Recension*. It seems that in the Old Kingdom these spells of rebirth and the promotion to an eternal afterlife were reserved solely for the benefit of pharaohs. However there is little reason to doubt that the spells had evolved from much older predynastic concepts. Chapter 130, for example, was a hymn to be recited on the birthday of Osiris. It was supposedly found in a cave which Horus made for his father Osiris – which Wallis Budge believed to show that it existed even under the First Dynasty.

During the New Kingdom (from 1567-1085 BC) the chapters and spells began to be painted on sarcophagi together with colourful illustrations; and hence are known as Coffin texts. The afterlife was now available to any Egyptian who could afford a lavish burial. Termed the *Theban Recension* it is from this era that we find the earliest and most authentic versions of the painted scenes. Sometimes a papyrus scroll of prayers and spells written in the short-form hieratic script would also be buried along with the deceased. The chapters seem to have had no particular order, as if they were being used in some instances as mere decorative art to fill the available space. To our modern eyes it all seems just so much gobbledygook and must surely have seemed so to many Egyptians.

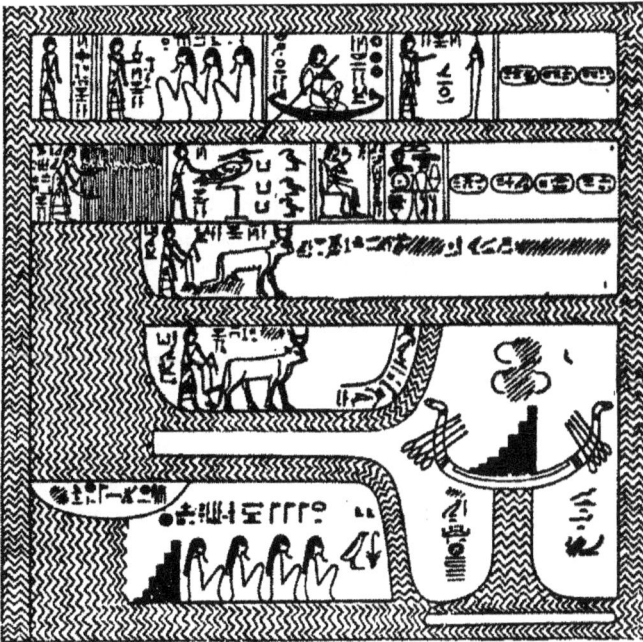

Figure 2.5 The Elysian Fields from the papyrus of Nebseni (reproduced by courtesy of the Trustees of the British Museum)

During the religious revival that accompanied the Saite Dynasty the funerary traditions were revised yet again to form the *Saite Recension* of the Book of the Dead. We now see the papyri in their most stylised and simplified forms, with standardised text and less use of colour. The sequence of the various chapters also became more regular compared to the haphazard usage of former times.

The spells would offer the deceased an opportunity for rebirth into an eternal hereafter free from toil, but first their soul would be assessed according to the conduct of their earthly life. Some chapters were more important than others. Chapter 1 is standard to all; concerning the Coming forth by Day, a hymn to the Sun-God Re in which the spirit is re-awoken into the afterlife; he finds himself in the Land of Manu – the Land of

Vindication; and the Manu Mountain where the sun sets, lies before him. Other chapters giving more detail on this reawakening are most common, but their inclusion seems to have depended upon individual choice and means. There were standard spells concerning the questioning of the deceased to determine if they are worthy to enter the Netherworld. Osiris is asked to weigh the deceased's heart and to pronounce that he is righteous; other spells gave the deceased protection from the various hazards, such as the evil serpent Apep; and also unleashed the *shabti* to do work for him and thus free him from eternal toil. Spell 13 for example asks: 'Prepare a path for me, that I may enter in peace into the beautiful West, for I belong to the lake of Horus…'

Some of the papyri describe the land of the dead in greater detail. The paradise of the worthy was called *Sekhet-Auru* or 'Field of Reeds' where the deceased would go to spend the afterlife. This was only a part of a larger divine agricultural region called *Sekhet-Hetepet* or 'Fields of Peace' where the company of Gods presided. These in turn were part of *Amenti*: namely the Underworld. Several papyri hold detailed maps of this place and they portray a rectangular region of tilled fields with irrigation canals marking the boundary and dividing the fields. The deceased is sometimes depicted driving oxen, reaping the wheat or rowing the canals in a boat. Sometimes rivers or streams are seen flowing through the fields always discharging themselves at bottom left of the scene, where lies a city, occupied by the host of gods. This format is most clearly seen in the earliest papyri dating from the Theban Recension. In the later equivalents from Saite and Ptolemaic tombs the rectangle is sometimes rotated and the waterways are reduced to thin blue lines, but they are recognisably the same place. We should assume that the oldest form from the Theban Recension preserves the most authentic form.

Some of the finest examples of these papyri depicting the Field of Reeds are on display in museums around the world. In the British museum are the Papyrus of Anhai (20th Dynasty: c.1100 BC); the Papyrus of the Nebseni (18th Dynasty c.1400 BC); and the Papyrus of Userhat (18th Dynasty: c.1400 BC) as were described by Wallis Budge. These show best the rectangular form with three or four sectors divided by canals. The lower sector is dissected by rivers and show the island described as the birthplace of the gods. A typical Nile boat is shown moored at the end of a canal. The Papyrus of Nebseni shows most clearly that the rivers discharge into the sea around the island. However, another earlier papyrus, the Papyrus of Ani (19th Dynasty c1250 BC) has the rectangular plain rotated east-west and this time with the rivers and island of the gods at lower right.

In the Paris Louvre is the Papyrus of Djedhor (Ptolemaic era) which shows the rectangular plain again rotated; and in the Turin museum, the Papyrus of Kha (Ptolemaic period) and in other later versions such as the Papyrus of Hori (Ptolemaic: 2nd Century BC). It would seem that the Saite Recension had standardised the fields as an east-west rectangle and the relevance of the older north-south layout had been forgotten.

The Greeks easily recognised and equated their own *Elysium* or *Elysian Fields* with the Egyptian equivalent. We may note the obvious similarity between the Elysian Fields of the Egyptians and the rectangular plain and its fields that Plato derived from Solon, which was quoted in the previous chapter. He described a similar rectangular plain of 3:2 proportions divided by canals with the longer side being north-south and the shorter dimension across the city; the city is located on an island at bottom left (or south-west) of the rectangle, where the rivers discharge. This detail is quite specific and it supplies more geographical information than is needed for a mere fiction. It is another of those mythological 'fossils'; the likelihood that such parallel geographical depictions derive from independent sources seems too much of a coincidence to accept. To make this clear, recall again the precise words used by Plato:

> ...but the country immediately about and surrounding the city was a level plain...it was smooth and even, and of oblong shape, extending in one direction three thousand stadia, but across the centre island it was two thousand stadia. This part of the island looked towards the south...

Figure 2.6 The Elysian Fields from the papyrus of Anhai
(reproduced by courtesy of the Trustees of the British Museum)

It would seem that a real location, which was related by Sonchis to Solon as a pseudo-history, had become absorbed into the religious mythology as part of the Underworld; the place to which the righteous would go to spend eternity alongside the gods.

We may see in this a belief that came into Egypt along with the Libyan Amazons and their goddess Neit; and it was recorded in her temple records as the place in the Atlantic where their gods were born, or we may prefer to say: where deified kings and demi-gods who claimed descent from gods had formerly lived. Their colonists then brought these beliefs from their homeland with them into the Mediterranean, just as later European colonists to the Americas would bring their own religion and remember their country of origin. However, after this homeland was lost in a catastrophe, the memories began to degrade, first to timeless legends, then to myth, and in later Egypt they would be absorbed as just another story alongside the African and Asiatic beliefs that composed Egyptian religion. But with the help of the few clues hiding in the myths we may be able to recover some of that lost history.

That Diodorus Siculus also knew of the Book of the Dead mythology is attested by his own words; and such quotations are important here in order that these concepts should not be dismissed as mere speculation of the present author.

> Orpheus, for instance, brought [to Greece] from Egypt most of his mystic ceremonies, the orgiastic rites and...his experiences in Hades. For the rite of Osiris is the same as that of Dionysus and that of Isis is very similar to that of Demeter, the names alone having been interchanged; and the punishments in Hades of the unrighteous, the Fields of the Righteous, and...were all introduced by Orpheus in imitation of the Egyptian funeral customs... [There follows a quotation from Homer's Odyssey 24, 1-2 and 11-14][12]

In the Egyptian myth of Osiris, the body of the murdered king was cut into many pieces by Set, only to be reassembled and reanimated by the goddess Isis. From their subsequent union was born Horus, to become one with the next king.[13] It may be that this myth was re-enacted in certain rites and ceremonies that some considered improper.

Diodorus goes on to describe how the Egyptians had localised the funerary myth in common usage, referring it to the River Nile rather than Oceanus and the 'meadows' west of Memphis had become the location of the Fields of the Righteous; and how other myths about Hades as known to the

Greeks also had their counterparts in Egypt:

> For the boat which receives the bodies is called 'baris'
> [Greek = a boat] and the fee is given to the boatman,
> who in Egyptian tongue is called 'Charon'. And near
> these regions they say, are also the 'Shades'...

This idea that the souls needed to be ferried across to the fields
is all that preserves the concept that they were originally
situated on an island. This is not specifically stated in any of
the extant papyri, but the equivalence of the Fields of the
Righteous with the Greeks' own Elysian Fields restores this
missing part of the mythology; these we are told were situated
on the Isles of the Blessed which lay in the furthest west.

In Greek mythology Elysium was a paradise where the
honoured dead were granted immortality. It was said to lie
somewhere in the west at the end of the earth alongside the
stream of Oceanus. We encounter it most memorably in
Homer's Odyssey, where King Menelaus is told that when he
falls, he will join the immortals, who reside at the world's end,
where the gentle westerly breezes from Oceanus blow over the
Elysian Plain.[14]

For the Greeks Oceanus was a great river that flowed right
round the known world, hence derives our modern word. In
Pindar's Odes Elysium was said to lie on the Island of the
Blessed.[15] By the time that this was absorbed into Roman
mythology it had become merely a division of Hades, where
the honoured dead might reside, separate from less favoured
shades. However, nowhere in Greek or Roman mythology do
we find the precise description of rectangular fields and rivers –
we are not told the shape or dimensions of paradise! These
details come only from Egypt.

We now see now why the temple of Neit at Saïs also held a
shrine to Osiris (of whom Herodotus declined to speak) and
why we should be confident that the temple there must have
held the images and spells from the Chapters of Coming Forth
by Day - including the maps of the rectangular Elysian Fields;
and that the Saite priests alone had preserved the historical
origins of the myth in a manner more authentic than elsewhere.

Figure 2.7 The Elysian Fields from the papyrus of Ani
(reproduced by courtesy of the Trustees of the British Museum)

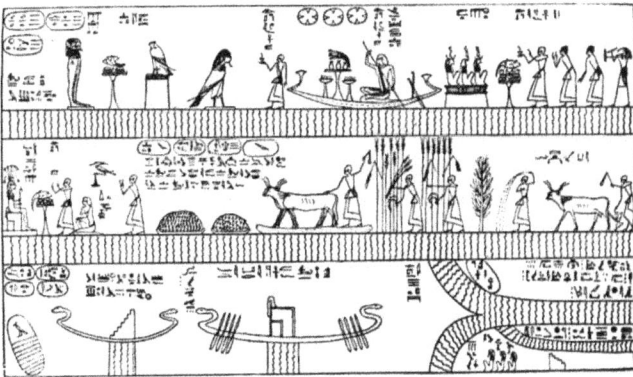

Figure 2.8 The Elysian Fields from the Turin papyrus (Turin Museum)

We may be sure too that Egypt had its own equivalent of academic specialists. It may be that Sonchis was one of only a few who could read the ancient records at Saïs and had made a study of them. As to when the knowledge and the records were lost there are several possibilities, of which the most likely is that they were destroyed during the Persian invasion. Herodotus, as we have seen, knew nothing of them when he visited the restored temple; he says that he spoke to the scribe there who kept the register of Athene's treasures. Undecipherable records may yet have survived until the neglect of all the temples in Roman and Christian eras that followed. The methods by which history has been lost to us are many and all of them a tragedy.

Amazons and Atlantian Colonists

Egyptian chronology and its kings is not the primary subject here and such things can be left to specialists to argue and await a suitably eminent voice to pronounce. The important lesson is that nothing is yet proven about the origins of Egypt. Here we follow the legendary accounts rather than artefacts from the ground.

From two ancient sources that should be respected we have accounts of a pre-dynastic invasion of Egypt from the western Mediterranean. One is supplied by Plato via Solon and recalls a campaign by an empire from the Atlantic, which had colonised the Mediterranean coast; the other coming via Diodorus Siculus and Dionysius 'Skytobrachion' is the campaign by the Amazons who claimed to have conquered Atlantians and were perhaps themselves earlier colonists from the west. Both stories may be seen to derive from unrelated historical records in Egypt. Neither history agrees with that which the Egyptologists derive for this period from the artefacts - so perhaps they would like to consider whether they are neglecting evidence of two completely different invasions of predynastic Egypt from the west, or are they dismissing just the one?

We may suppose that the 'missing' dialogue of Hermocrates was intended to describe in more detail the Atlantian war with Athens and this might have been revealed as the campaign of Myrina. Consider the similarities. Both were campaigns in which an army from western Libya threatened Egypt and swept on round the Mediterranean coast to the Aegean seeking to conquer the entire ring of the eastern Mediterranean. However there are differences: Plato's version describes only kings of Atlantis and no mention of a Queen; that of Diodorus says that Amazons (perhaps in alliance with Atlantian colonists) were defeated in the Aegean region, not by Athenians but rather by an 'alliance' led by a Thracian. Plato also says Athens was part of an alliance – but that these all deserted her, leaving Athens to fight alone. Could these two differing accounts both be a degraded memory of the same campaign of conquest?

We are told by Diodorus that Queen Myrina came to an alliance with 'Horus son of Isis' who ruled Egypt and then continued on across the Nile into Asia. Perhaps here we see here not the mythical god Horus, but king Horus-Aha, or one of the later kings of the First Dynasty who called themselves by a Horus title. Myrina was not interested in the Nile valley. She wanted to control the circle of Mediterranean coast from her homeland: Sicily-Malta-Libya and right round the Levant; she didn't want to head-off down the Nile. So she came to an alliance, we may call it a marriage, whereby she, via one of her daughters or sisters, ruled the Nile Delta while the king of Upper Egypt avoided a war and kept his throne in a kind of dual-monarchy. We may recall the exact form of words used by Diodorus: 'she struck a treaty of friendship with Horus the son of Isis, who was king of Egypt at that time'. Here perhaps we see Horus-Aha – a real historical king – caught in that moment where history is degrading to legend and his story is becoming amalgamated with the god-Horus so to be dismissed as a figure of myth.

Myrina too, if her name has come from Egypt, may be a degraded memory of a real queen, perhaps named Mer-i-Nar or Mer-i-Neit. With Neit as a component of her name she too came to be merged with the goddess as a single figure of myth. The story says that Myrina founded many cities and therefore it is likely that Saïs in the Delta was one of these locations. It was the Libyan Amazons who recorded there the story of the home of the Libyan gods on an Atlantic island. All of these events we may place in the period just before Dynastic Egypt began.

We also have a historical account of a catastrophic event of abnormal proportions that occurred in the late predynastic period just before 3100 BC. Egypt is the only civilization (save perhaps for Mesopotamia) where we have anything like a historical chronology to which we can attach these events; and we have the opportunity to confirm it by archaeology. Elsewhere in the world we have only oral tradition and legends, to compare with radiocarbon dates from artefacts. It is unusual for archaeologists to consider the climate and sea-level evidence contemporary with their artefacts. We know also that the late predynastic was a period when the formerly green Sahara region rapidly turned to desert, driving the indigenous tribes to seek refuge in the Nile valley and along the coasts. From the sources above we see that the *emergence* of the Nile delta five thousand years ago was contemporary with earthquakes in the Aegean, yet *submergence* along the western Mediterranean and Atlantic coasts.

From Plato, Diodorus and perhaps Herodotus also, we see a belief among the Libyan tribes and in Egypt that before the catastrophe there had been a powerful empire along the Atlantic coast. This empire had sent colonies and conquests into the Mediterranean. Quite suddenly, the empire that had been a threat on their borders just disappeared in a way that they could not understand.

The Egyptologists' Fear of Atlantis

However, we must not stray too far beyond the chain of evidence and coincidences set out above. If you are one of those who would still dismiss ancient myths and prefer to treat Plato's dialogues as mere fiction and fable then consider the following. In composing his supposed fiction, how did Plato know that modern Egyptologists would find the First Dynasty of Egypt and the Twenty-sixth Dynasty to be the correct eras for the religion of Neit to be at its most prominent? How could he have known that the Egyptian Book of the Dead was also revised at that era; and that its chapters contained descriptions of the rectangular Elysian Fields with dimensions in the ratio 3:2 showing canals and internal features laid out just as Solon recorded? He could not have known of the synchronism with Myrina and the Amazons of Libya, or of the Nar-mer / Mer-i-nar palette and the other archaeological clues to a First Dynasty date for Neit's temple. These are the details, the 'fossils', in the story that lend it authenticity. It is reprehensible that Egyptologists and other academics have, for so long, neglected these evident coincidences of fact.

Notes and References

[1] Book 1.43.5-44; Oldfather's translation

[2] Book 1.25. 6-26; Oldfather's translation

[3] This is Birket el Qarun in the Fayum Depression, which was formerly much larger than today.

[4] journals.plos.org/plosone/article?id=10.1371/journal.pone. 0069195, ref 37

[5] Emery W.B. Archaic Egypt, Penguin, 1961 p52

[6] I am aware that I am not the first to suggest this, but I am unable to trace now the reference where I saw this thirty years ago.

[7] Hoffman p332 citing Emery 1961 p47

[8] One may note how similar in format this early inscription is to the later Book of the Dead illustrations showing boats sailing on a canal.

[9] The only Egyptologist I can find to cite in support of a similar opinion would be Schulman; see *Schulman, AR (1991–92), Narmer and the Unification: A Revisionist View, Bulletin of the Egyptological Seminar, 11: 79–105.*

[10] *Anđelković, B (1995), The Relations Between Early Bronze Age I Canaanites and Upper Egyptians, Belgrade: Faculty of Philosophy, Center for archaeological Research*, p 31

[11] Gardiner, Sir Alan, *Egypt of the Pharaohs*, Clarendon, Oxford, 1961, pp366-7

[12] Diodorus Siculus, I, 96

[13] Plutarch, Isis and Osiris, 356, 12

[14] Odyssey, IV,549-643

[15] Pindar, Olympian Odes, II, 70-75.

3

In a Single Day and a Night

How could an island disappear beneath the sea in 'a single day
and night'? The clue is there in Plato's own words - a single
day and night is one rotation of the Earth. This is another of
those mythological fossils that we can attempt to probe with
science. If Plato's story were a mere fiction then we must ask
again why he needed to include such a specific detail. To say
that submergence occurred rapidly suffices for the needs of a
story-teller. Neither does he give us any indication of
volcanism as the cause, or describe anything that we might
recognise as a tsunami. The island (or at least a part of it)
simply 'disappeared beneath the sea' in a manner that could not
be explained or understood – all within a day.

Over the years so many ideas have been put forward to
explain the nature of such a cataclysm, some genuinely offered
– others so absurd that it is hard to accept that even their
authors ever believed them. Before modern oceanography ruled
out the existence of mid-ocean land-bridge continents (Mu,
Lemuria, Atlantis, etc) and before plate tectonics was properly
understood, it was perhaps permissible for Donnelly and others
to propose that an island as big as a continent could descend
beneath the waves in the manner inferred by Plato. Yet there
are *real* examples of submerged land-masses – for example the
two islands of New Zealand are just a part of the ancient
Zealandia continental plate that remain above the sea. The
North European continental shelf surrounding Britain and
Ireland is another region where shallow seas have only recently
covered the continental crust.

Atlantis Goes on a World Tour

As an author on Catastrophism and related subjects, one is occasionally asked to interview on television, or to produce a book for the American market. Publishers and TV producers on ancient mystery subjects are interested in volume of sales, not in establishing facts. If the masses prefer to buy that ancient aliens arrived and established Atlantis on Earth then that is what they will put out! The market in North America is many times that of Britain and other English-speaking countries. They don't want to hear that a Neolithic civilisation may once have existed somewhere off the coast of Europe, or even in the Aegean; they would much prefer to move it to Bimini, or to Cuba, or even South America. This imaginative output, together with related fiction and pseudo-science based on outmoded 1950s theories, have together contributed to making the subject of catastrophist geology something of a no-go area for legitimate scientific investigation.

Foremost among the early authors was Ignatius Donnelly in 1882. In his era the proposal that a large island might have sunk in the middle of the Atlantic Ocean – right opposite Spain where Plato had located it – was acceptable science. It was plausible to theorise that 10,000 years ago colonists had sought refuge on both sides of the ocean, in America and Ireland. On reading his book today one can't help feeling that he made the best of it based on the scientific orthodoxy then prevailing, for example in his Chapter II he describes their Egyptian 'colony', where we find:

> The mythology of Greece was really a history of the kings of Atlantis. The Greek heaven was Atlantis...

In the nineteenth century no-one could possibly know what lay on the ocean floor. There was no reliable chronology for Egypt, or any way of dating Maya archaeology. He could not know that the Pyramids of Egypt were thousands of years older than those of Central America to which he compared them. Unfortunately, when modern geology swept away the concept of sunken continents and archaeologists revealed the true age of the civilizations on both sides of the Atlantic Ocean then the more sound parts of his hypothesis were swept away along with the rest.

One of the stranger notions popularised by more recent authors was that at the end of the Ice Age, a mid-Atlantic continent drifted-off to become Antarctica. No, don't laugh, a lot of people actually believed that one! The suggestion derives loosely from the theories of Charles Hapgood in the 1950s, before plate tectonics was fully understood. It was based on a medieval map by the explorer Piri Reis, which appeared to show a precocious knowledge of Antarctica. In fact he was just randomly depicting the 'great southern continent' that many early explorers expected to find.

The modern science of plate tectonics did not exist until the 1960s when advances in oceanography first revealed the volcanic mid-ocean ridges and sea-floor spreading. Land bridge continents were variously proposed to account for similar geology and fossils on either side of vast oceans. Continental drift, as it was then termed, was the new concept trying to replace the notion of sunken land-bridge continents. It could be seen that the shape of continental margins appeared very similar as if they had earlier fitted together and had drifted apart. This still left scope for some continental fragments to have been submerged over the millennia, perhaps in the Atlantic too.

Sonar mapping of the ocean floor revealed the sea floor spreading ridges and transform faults. It could now clearly be seen that the ocean floors had spread apart only gradually over millions of years due to volcanic action at the mid-ocean ridges, carrying the continental plates with them. The notion of rapidly 'floating' continents became an absurdity.

Pole Shifts and Ice Ages

The present author's published theories on this subject were set out fully in *Atlantis of the West* and *Under Ancient Skies*. However a summary will be needed before we can proceed. Use of specialist terminology will be kept to a minimum in order to retain the attention of the general reader; and evidence will only be cited in places where something more recent has arisen to counter or confirm the hypothesis. Other than a theory of pole shifts, there is nothing here that would not be considered as current orthodox science.

The weakness with many catastrophist theories, going right back to Donnelly, is that they neglect the fact that our planet is a rotating system, with all the limitations of real physics and astronomy that accompany that. The Earth's rotation is delicately balanced as it revolves about its axis of figure and the sea level is similarly aligned about that axis. A 'continent' could not just subside into the sea, or destroy itself volcanically, without causing rotational instability and worldwide consequences.

Posit that for some unspecified reason, a large land mass really could sink or float away across the crust; this would alter the figure of the Earth: the geoid. The figure axis (which must always pass through the centre of gravity) migrates to a new position. It then has to spiral about the axis of rotation until they can realign. This would certainly cause inundation around world coastlines and trigger earthquakes, but since this is all internal to the Earth, the angular momentum of the planet – the obliquity and the length of day – would scarcely be affected.

Approach this from the other direction. Suppose that an external force acts upon the Earth: a small energetic comet impact such as the ancient *Phaeton*, or a force as yet unknown to science, acts to change the Earth's angular momentum. This would affect all the characteristics of the diurnal rotation: obliquity (the axis 'tilt'), the 24-hour day, as well as displacing the geographical poles. In addition to affecting sea-levels worldwide it should leave evidence of climate change and have effects for the calendar. However, the weakness of this hypothesis is that the energy required to change the rotation is so enormous and 'earth-shattering' that specialists would surely have found the evidence.

One may posit that, as with the earthquakes we experience today, stresses stored in the crust after an earlier quake could be suddenly released. It would then be possible to hide the true causal event in an ancient impact event far back in geological time.[1] This might have left a section of crust (or even the core) trapped above its equilibrium height. Were this to slip suddenly then it could trigger a pole shift without evidence of a

59

contemporary astronomical cause. Again, the difficulty is that modern oceanography shows no such submerged continental fragment in the Atlantic Ocean; and if sea-level change along Atlantic coasts were consequent upon massive earthquakes or volcanic subsidence elsewhere then there should be synchronous evidence of giant tsunamis around world coastlines. One has to conclude that field geologists would have found such evidence.

Authors of 'ancient mystery' books and spin-off fiction would prefer to hide a vague geological upheaval thousands of years ago back at the end of the ice-age; or to localise and simplify it somehow. But here we must follow the *very specific* trail of evidence established in the previous chapters. That evidence directs us to a rapid submergence along the Atlantic coast of Europe just 5,000 years ago which also had wider geological effects within the Mediterranean. Its effects were not localised. If it were a real event then there must be other worldwide evidence to be found dating from the same era.

Ice Ages and Glacial Eustasy

The author's interest in the theory of pole shifts came about as part of research into the cause of ice ages and their sudden end; and dissatisfaction with the vague gradualist explanations and circular arguments that pervade the standard theory.

It is should not be necessary to give here a complete account of how knowledge of ancient glaciations developed through nineteenth century advances in the understanding of Earth prehistory; and the realisation that large ice sheets must formerly have covered large parts of northern Europe and Canada. The fossil record reveals alternating glacial and interglacial periods over the past few million years.

An obvious consequence of ice sheets that have melted away is that the melt water must have gone into uniformly raising the level of the ocean worldwide; and that some parts of the continental shelves that are now submerged must in earlier times have been above the sea. This is apparent in the North Sea, the Caribbean, Indonesia and Australia where low-lying coastal regions have been inundated. However, in northern regions such as Scandinavia and the Canadian arctic, land can be seen to have emerged since the melting of the ice. Early theories of pole shift were rejected by nineteenth century physicists and along with them went the possibility that such shifts might have caused the emergence of polar land. However, the concept of *glacial isostasy* was retained to account for the gradual rebound of the land due to the removal of the great weight of ice; and hence the term *glacial eustasy* for its effect on the sea level.

It cannot be denied therefore that many former coastlines now lie submerged offshore of modern coasts and that most sea-level rise since the ice age can be explained this way. Field researchers would seek areas of the crust where local earth movements could be ruled out and search for dating evidence to confirm the history of ice-cap melting and re-advances. Usually they would find what they were seeking. However, in some places around the coasts they would find the opposite: an anomalous raised shoreline dating to an era when theory dictated that it should have been submerged. The concept of isostasy was then employed to explain a rise of land locally, thus hiding the expected worldwide trend. In other words, the glacio-eustatic theory must be correct, but was being masked by a local anomaly. This is the problem you will often find if you read dozens (hundreds) of research papers and try to discern a pattern.[2] The field specialist has to say what is expected in order to get their paper past an academic referee; it is then available for others to cite and thus becomes embedded in the textbooks as proven fact. It becomes 'official' science.

From the preceding chapters you may realise that the present author, acting as a kind of 'Liverpool jury' could not convict the defendant on such dubious evidence. One cannot claim that all sea level change is due solely to episodes of melting and freezing of ice sheets, and then try to smuggle the cause back to gradual processes, when proof of that theory depends crucially upon sections of the crust being allowed to bob up and down for no apparent reason. To the sceptic these are indications that something is wrong with the basic theory.

Another inconsistency was that Arctic Russia and Siberia, today the coldest regions of the northern hemisphere, were never glaciated. In these regions hairy mammoths and mastodons were able to survive on the 'Mammoth Steppe' eating grass and wild flowers; while at the same latitude tundra and deep ice sheets covered large parts of what are now Canada and the USA. One is expected to suspend the basic principles of climate geography taught in every school in order to embrace Ice Age theory.

A further weakness is that the episodic nature of ice advances and retreats around the world must be explained. The Milankovich theory of long-term variations in the Earth's orbit due to the pull of sun, moon and planets came to the rescue – so long as these astronomical arguments are kept suitably vague and gradual. For older epochs such climate changes can be passed-off as a gradual process, but the closer we approach to the present day the less convincing this becomes. Recent organic matter can be accurately dated by carbon-14. In some places climate transitions could be seen to have taken place rapidly within the space of just decades.[3]

The relatively warm period in which we now live is termed the *Holocene*, to distinguish it from the *Pleistocene* epoch – the age of ice. The transition event termed the *Younger Dryas* is deemed to have commenced around 12900 BP as a short fluctuation in the perceived gradual warming since the last glacial maximum around 24000 BP. Following this brief cool oscillation, our present warm 'interglacial' begins, into which all of human civilisation falls. We are concerned here primarily with Europe and this is the period when the ice sheets finally retreated from Britain, Ireland and Scandinavia, bringing the landscape that we see today.

Climate scientists and geomorphologists will therefore estimate how much ice has melted, based on the weight of ice required to depress the crust to the extent suggested by raised ancient shorelines. Sea level researchers will then take these ice caps as a starting point and use that to calculate how much the sea level must have risen when this ice melted. However, if the apparent rise of polar land were due to some other contributory cause, such as adjustment of the geoid consequent upon a pole shift, then such models would be invalid; but the weight of text books says that pole shifts are not orthodox science. It doesn't have to be an *either-or* case and each pole shift need be only a tiny fraction of a degree. If some of the climate fluctuations were caused by pole shifts then any evidence of the former shorelines would now be eroded or submerged deep offshore – except perhaps for the most recent sea level changes dating from the Holocene.

The Four Quarter-spheres Pattern

Sea level research has mirrored the investigation of ancient climate. Researchers would expect to find the sea level rise due to melting ice and regression due to cooling when the ice sheets were growing. This pattern should give a consistent world-wide sequence of raised beaches and sunken features for any particular date, but it has been apparent since the work of Rhodes Fairbridge in the 1960s that the search for such a consistent worldwide sea level curve is fruitless; and that at a local level, the evidence is much more confusing.

Taking 5000 BP (c. 3000 BC) as a focus, we may see that there was emergence at the Nile Delta at the same time as submergence elsewhere; and yet this was contemporary with a warming of the desert climate rather than a cooling. A 'line of neutrality' is apparent running down from Denmark, through Italy and on through Africa. To the west of this (in the northern hemisphere) we find evidence of submerged coastlines while to the east of it there was emergence or rising land. This apparent uplift of land continues from Scandinavia to Mesopotamia, across central Asia and as far as China and Japan.

Correspondences in the opposite southern hemisphere quarter-spheres are more difficult to detect as there are fewer southern coastlines to examine. Nevertheless, we may see raised beaches from this era along the coasts of Patagonia and (with less certainty due to earthquakes) along the Chilean coast; while in the opposite quarter-sphere we find submergence around Australian coasts, where the shallow shelf around the Great Barrier Reef has been inundated since this date.[4] Aboriginal folk-memory remembers the formation of the islands. Wherever we see 'raised beaches', in Argentina or elsewhere, this requires a (presumed) tectonic lowering of the coast, then a standstill to cut the beach, followed by a fall of sea level, or uplift, to expose the raised beach – but through all this time the sea is predicted to have been rising worldwide due to the melting of ice at the poles; and tectonic forces should be raising the Andes mountains not the opposite. You see the complexity of the problem!

In central Asia we may note the reduction that has occurred in the extent of the Aral and Caspian basins, even before recent Soviet-era damming of the rivers. An examination of the descriptions of Herodotus and the campaigns of Alexander the Great reveal that these rivers were formerly much more extensive; great rivers that in earlier times flowed to replenish the inland seas now dwindle and evaporate in the desert. This is suggestive of a relative uplift or tilt of the land in central Asia in recent millennia.

Further south in Mesopotamia and also in China we see a growth in the major river deltas since the rise of the earliest civilisations, just as is noted in Egypt. In Iraq, ancient cities such as Ur which were once at the mouths of their rivers are now found inland of the delta. In China, the earliest farming civilization only colonised the delta of the Hwang-Ho River since five thousand years ago – around the same time as this was happening in the Nile Delta.

The author's published research based on an examination of sea-level curves and other evidence for 5000 BP would suggested that there was a worldwide pattern of correspondence in alternate quarter-spheres as would be expected for a pole shift.[5] To clarify this further: in the direction of a pole shift, there should be a quarter-sphere of lowered sea level with a corresponding rise of sea level behind it. In the southern hemisphere there should be the opposite pattern. Overall, the pattern for 5000 BP would be consistent with a small shift of the North Pole (no more than a quarter or a third of a degree of latitude) from the direction of Ellesmere Island to its current location.

The quarter-sphere pattern would actually suggest that the deepest coastal submergence (before considering any adjustment at the mid-ocean ridges) could actually have occurred off the east coast of North America and the Caribbean. However, investigation of whether any submerged archaeology might lie in that region is beyond our scope here.

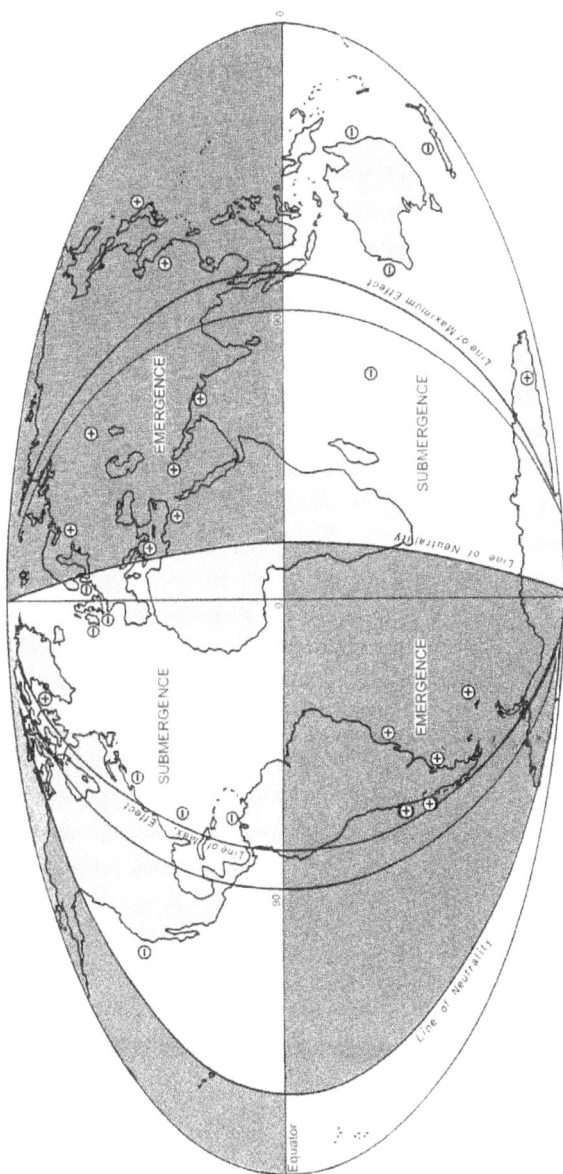

Figure 3.1 The four quarter-spheres of sea-level change around 3100 BC. This map was originally included in *The Atlantis Researches* (1995)

As the saying goes:

In fourteen hundred and ninety-two
Columbus sailed the ocean blue...

Before that date Spain knew nothing about Mexico; and Mexico knew nothing about Spain. Any submerged civilization that might predate the Olmecs and the Mayans would be entirely independent of old world evolution. Anyone who would seek evidence of a five-thousand-year-old lost civilization in the Caribbean region should look to the legends left by the Mayans rather than those of Egypt or Greece.

Flood Myths – a Closely Related Pattern
An examination of world-wide myths of an ancient flood catastrophe will reveal an analogous pattern in alternate quarter-spheres as noted above. The Biblical flood is principally a near-eastern and Mediterranean phenomenon. However, we may find corresponding flood myths in the Americas and even among Australian Aborigines. However, in areas such as Finland, Asia and the Pacific where flood myths are weak we find instead memories of rising land, or of islands emerging from the sea.

Although such myths are quite timeless and impossible to accurately date, it is worth noting again the way that myths evolve from decayed history. There has probably always been a flood story among human myths of origin, recording, as Sonchis of Saïs related, the many natural disasters that have occurred – but the Greeks recalled only one.[6] Events and characters associated with the most recent earthquake or tsunami therefore become conflated with older memories to form a single myth of a 'great flood'; but the detail of the most recent experience should be dominant. We should just note, that some flood stories are 'inundations' – the water comes from below; while others are 'deluges' – the water comes from above.

To give just one example of this: in 1998 William Ryan & Walter Pitman published their research into the flooding of the Black Sea basin. Their theory would relate the Mesopotamian and Greek memories of a 'great flood' with the rapid entry of the sea via the Bosphorus Strait to join with the freshwater lake (the Euxine Lake) that had formerly existed there. This was not entirely a new theory as Bulgarian researchers had proposed something similar in the 1970s. Later exploration and dating would suggest that this inundation took place around 7400 BC, somewhat earlier than the 5600 BC originally suggested. Of the discussion that has arisen since Ryan and Pitman's publication of their theory, no-one seems to have suggested a geodetic or pole shift explanation for such a flood.

However, archaeologists would prefer to link the Biblical flood (the 'Deluge') with memories of localised river floods in the Tigris-Euphrates basin that are recorded in the earliest Sumerian king lists; and which date, once again, to the years just before 3100 BC. As with the Egyptian chronology this dating was achieved by counting back the reigns in the Mesopotamian kings lists from the earliest reliable fixed point. So how are these two disparate pieces of evidence to be related?

It may well be, as some suggest, that earlier flooding of the low-lying land around the Euxine Lake did cause a dispersal of the tribes formerly living around its shores, each taking with them a flood story. In the various regions where these peoples then settled, earlier or later 'flood-hero' figures, such as the Greek: Deucalion, Welsh: Nevydd; Babylonian: Utnapishtim; or the Biblical Noah, became absorbed into the local variant of the story.

However, we must resist the temptation to pull on interesting threads that distract us from what was happening along Atlantic coasts five thousand years ago.

The Mid-Holocene 'Golden Age'

Climate change during the Holocene has been far from uniform. Climate scientists recognise a period of steady warming climate from about 9000 BP to 5000 BP. The period just before 5000 BP (or 3000 BC) also corresponds to the height of the mid-Holocene warm period. In Europe this warm period is known as the *Atlantic* pollen zone – but for correspondence with other parts of the world you will find the same peak climate regime termed as the Hypsithermal, Altithermal, Holocene Megathermal, etc. To be more precise, it was a period of warmer winters and less extreme temperatures.

Probably, the only task that could be more frustrating than to search for a pattern in published sea-level curves would be to consider the contradictory climate research covering the same period. An examination of the available published climate evidence will reveal the typical language used by researchers to make this all fit within the expected pattern of gradual change; consequent upon the same orbital cycles as are cited to explain climate swings during the age of ice. Where the evidence fails to fit the prediction one may see attempts to explain non-compliant data as, for example: 'ice-albedo forcing', 'orbital forcing', and suchlike jargon phrases; which are unconvincing and should not convince us of anything except that the specialists are not really sure what the cause was.

Building upon the pioneering work of Iversen in Denmark the 1950s, a series of pollen zones for Europe were established.[7] These derive mainly from cores taken in marshes and peat bogs and they reveal the post ice-age transition from tundra to pine forests and on to temperate deciduous forests. Improvements in dating techniques have served to confirm rather than challenge these basic definitions.

The Holocene opens with pollen evidence of birch forest growth on the newly ice-free northern landscape, overtaken by the slower growth of pine forests up to about 5500 BC; hence it is termed the *Boreal* period (c.7700-5500 BC). Further warming to the post-glacial maximum is then indicated by the spread of deciduous forests of oak and elm, further north than these species are currently found, as the warmer *Atlantic* period took hold in Europe. The Boreal period corresponds to the *Mesolithic* era as defined by the archaeologists – supposedly a time of tribal hunter gatherers. It is also the era (so the theory suggests) when the North Sea was dry and Britain was still linked to Europe as the melting of ice sheets continued to raise world sea level. The warm Atlantic phase corresponds loosely to the *Neolithic* (or new stone age) of Europe, by which time the sea level rise had drowned the European continental shelf and isolated Britain from the European mainland; it is only at this era that the archaeologists detect the first indicators of settled farming in western Europe.

Also of relevance at this era was a phenomenon known as the *elm-decline*. Much debated over the years by climatologists and archaeologists. Elm began to dominate the north European forests as the climate became milder, but elm trees cannot tolerate cold winters. Researchers find a decline in the pollen of elm and related species, which seems to correspond with human forest clearance, but the precise cause, remains elusive. Modern radiocarbon dating has suggested that the decline began around 4940 BC and ended around 3430 BC after which the forests began to regenerate. The most convincing cause so far advanced is human felling of mature elms together with selective feeding of domesticated animals on the nutritious young elm shoots.

Such then is the situation up to our period of interest around 3100 BC. The pollen record then shows a sharp decline in the northern range of deciduous trees, particularly of elm. This would be indicative of return to a cooler climate and this period has therefore been termed the *sub-boreal* (3000-500 BC). For archaeologists this climate is the background to the Late Neolithic, Bronze Age and early Iron Age of human activity. Such dates of course, should be regarded as approximations, but the transitions are too abrupt to be passed-off as solely due to orbital factors.

The transition corresponds with the so-called *Piora Oscillation* (3200-2900 BC) named after the Alpine valley where it was first identified; an apparent cold snap lasting about two hundred years. At this time, Alpine glaciers re-advanced down the slopes as the equable Atlantic climate apparently went into reverse. It should be noted however, that the same event that could cause fluctuating sea-levels should have a corresponding effect on the snow-line, which is simply a height above sea level.

In other parts of the world the mid-Holocene warm period is also found, but correspondence with the other European pollen zones is disputed. It is worthwhile to look more closely at what was happening around the world during the climatic optimum. As already noted, in Egypt the period up to 3000 BC is the predynastic and early Dynastic. In the Sahara the climate was wetter than today and a grassy steppe landscape was occupied by typical African species. Further south, Lake Chad then formed an inland sea occupying what is now the dry Bodélé basin; and seasonal rivers flowed into it – but this shrank to a greatly reduced size within a few hundred years of 5000 BP.[8] The transition to desert conditions appears to have set in very rapidly.

Herodotus, who gives us our only historical glimpse of North Africa, reveals that the desiccation process was continuing as late as 500 BC.[9] He describes the pastoral farmers who lived around the desert oases: Ammonians, Nasamonians and Garamantes – whom he says were numerous and had learned to farm the oases by spreading soil over the desert salt. One tribe, the Psylli, he says (citing a Libyan legend) had given up when their water supply failed; and the entire tribe had simply wandered into the desert and let themselves be covered with sand!

Further west from modern Tunisia, he placed the Atarantes who cursed the rising sun for burning their land; and even further west he describes the Atlantes whom he says were named after the mountain. We cannot be sure that the Atlantes of Herodotus were the same people as those Atlantians whom Diodorus described as possessing 'great cities' at an earlier time – i.e. before 3000 BC in this context; but it would indicate that their fortunes had greatly worsened since that era. In Roman times Pliny would describe 'the Atlas tribe' as reduced to 'below the level of human civilisation'.[10] We are fortunate to have the account of Herodotus and it shows us how much historical detail we are lacking in other regions such as Europe, where we have only archaeology to tell of the existence of ancient peoples. The histories of Herodotus cover almost exactly the time around 500 BC that climate scientists define as the transition between the *Sub-Boreal* pollen zone and the *Sub-Atlantic* in which we currently live.

Perhaps, of greatest value for those of us who would study mythology and legend, is that this period of warm equable climate: the 'golden age', may be discerned as a background to many of the myths. Such references allow us to assign such stories to a loose chronology and so we can build a pseudo-historical sequence for when the events must have taken place, during or after this warm phase. It offers an opportunity to assign some of the catastrophic events in myths and legends to a real physical occurrence that can be explored by scientific method.

Tree Rings and Ice Cores

A revolution took place in archaeology between the 1950s and the 1970s. The first of these was the adoption of carbon-14 dating to verify the age of ancient artefacts. Up to that era, the only method available to archaeologists was 'cross-dating', whereby pottery and weapon styles could be linked back to the Egyptian king lists by the discovery of Egyptian artefacts at archaeological sites; and then such linkage carried further away from Egypt by a similar process. With hindsight, we may see that this didn't work very well. The discovery of carbon-14 swept away this chronology and cultures in Europe that had been confidently dated to the second millennium BC had to be revised a thousand years earlier. For example, Stonehenge in England was formerly believed to have been built around 1800 BC consequent upon diffusion of civilised culture from the near east – but when radiocarbon dating became available it proved to be older than the Pyramids.

However carbon-14 dating has many uncertainties. It is dependant upon the ratio of 14C isotope present in the atmosphere when it was taken-up by the living matter. The 14C dates also suggested anomalies in the long-established Egyptian chronology that Egyptologists simply could not accept. This variation became further apparent when the method was applied to the annual growth-rings of long-lived trees; these can be counted back from a known date to give a precise annual chronology that can be regarded as calendar-year equivalent.

The tree ring dating reveals that there was an excess amount of carbon-14 in the atmosphere before about 1000 BC making samples appear younger, leading to an error of as much as 700 years by 3000 BC. There is some evidence that before this time the level of carbon-14 in the atmosphere was more constant, as it is today. Archaeological certainties were again thrown into doubt. Specialists therefore began a careful analysis of the annual growth rings of ancient trees in order to calibrate radiocarbon dates.

From the 1970's these tree-ring corrections began to be applied to published radiocarbon dates in archaeological research and also for climate and sea level data. This calibration has allowed the historical dates for Egyptian kings to fall back into line with their expected range – although at some periods the correction curve can offer more than one possible date. The reasons for these fluctuations are complex and we need not introduce yet further complication here. It is safe to say that nothing is yet proven about the cause of variations in atmospheric carbon-14.

A further discovery associated with tree-ring dating was that certain years show thin growth-rings, indicating reduced spring and summer growth due to a climate fluctuation. The sequence derived from Irish bog-oaks reveal the best data for Europe. At 3199 BC this shows a group of thin rings over a roughly 20-year period. One point to explore here is: what would happen if there were zero growth in a particular year? What if there simply were no summer and the trees conserved all their energy?

Similar low growth events are found at 1153 BC and earlier at 4377 BC. Another event at 1626-8 BC has been variously linked with the dust veil caused by the Santorini eruption – though as usual there is no agreement among the specialists as to its precise dating. A more recent low-growth event around AD 535 has been attributed to a dust veil from a Central American volcano or perhaps from Krakatau; and so the general assumption is that all the earlier events record a volcanic eruption.

Correspondence may be found in cores taken from ice sheets. A core from Camp Century in Greenland reveals high acidity at 3250 BC. This might be regarded as recording a quite different volcanic event from that at 3199 BC in the tree rings sequence – if it were not that similar sulphate acidity events are also recorded at 1100 BC and 4400 BC. The reason why the ice-core events do not correspond exactly with the tree-ring sequence is one for the specialists to resolve. However, it seems more likely that melting in a particular period could lose a period of years from the core and therefore the tree ring chronology is the more trustworthy indicator; for example, another ice-core from northern Greenland reveals that the ice sheet completely melted away during the mid Holocene warm period and has re-established only in the last four thousand years.[11] Overall it would seem that the tree-ring date of 3199 BC gives the best candidate to record the true date of an extraordinary event that may have triggered the other long-term climate and sea-level changes.

So would it be possible to prove that the fluctuations in the tree-ring record and ice cores were caused by wobbles of the Earth's rotation triggering seasonal variations and volcanism? One has to say 'no', but – *and it is an important 'but'* – if such a phenomenon were to occur then it should reveal itself in low growth events in tree-ring sequences and as melting events in ice cores; and the evidence should look much the same as is observed. The signature of a disturbed rotation is there if one wishes to interpret it. It is just that pole-shifts and the related wobbles of the rotation are not approved science and the specialists are not primed to notice such evidence.

Pole Tides and the Earth's Wobble

In discussing the subject of pole shifts and rotational wobbles, it is important to appreciate that the permanent axis of rotation cannot simply migrate from one stable position to a new stable position, regardless of what may be the cause. Before stability can be restored there must be a wobble phase lasting around twenty years; and also a second mode of much longer duration, perhaps persisting for thousands of years. Both modes must be triggered and occur together. This is not a simple subject to explain without mathematics and so a little more of the related geophysics and astronomy is summarised in Appendix B for those who wish to pursue it.

Suffice to summarise here that the short-lived wobble causes a spiralling of the rotation axis within the body of the Earth plus an accompanying small motion of the axis in space.

The other wobble is due to the presence of the fluid inner core and causes a similar spiralling motion, mainly in space with relatively little geographical pole shift. However it should persist for much longer – the geophysicists can't say precisely how long because it is barely detectable to modern instruments. One may think of it simplistically: that the former motion is rapidly damped by the elasticity of the crust while the latter is 'lubricated' by the fluid of the core. Both polar motions have a calculated period of around 13-14 months. Again one should stress that knowledge of the characteristics of the deep crust and mantle depend upon observation of these tiny motions on the stable Earth of today and no-one can be quite sure how the interior would behave under extreme conditions.

Only the short-lived motion would cause a *pole tide*. World sea level is a flattened spheroid, bulging at the equator due to the rotation. The solid Earth may be considered as a similar flattened spheroid with the continents projecting above it. To visualise the pole tide think of the two slightly different ellipsoids intersecting at the coastlines. Both are rotating together every 24-hour day about the same axis of symmetry. If the shape of the Earth should suddenly change then the figure axis would jump to a new position. The new figure axis then has to spiral about the rotation axis taking approximately 14 months to complete each circuit. This continues until the two axes can realign and the motion is 'damped'. During this period the ocean would alternate between spilling over the land and then retreating far out from the shore: this is the 'pole-tide'.

We have surely all seen on our television screens the destructive effects of the tsunamis in the Indian Ocean and Japan. These are caused by the long wavelength tsunami waves sweeping onshore carrying the energy of undersea earthquakes. It is important to visualise that the tidal bulge associated with a pole tide would not be at all like that. The tidal bulge of the lunar tide sweeps round the Earth every day and we note the sea going in and out of harbours, varying monthly with the moon, without causing massive destruction at the shoreline. The bulge of a pole tide would be analogous, but much higher and would vary over a 14-month period. However the tides would surge further inland and then retreat to expose more of the sea bed. This is the threat that could make an island disappear beneath the sea 'in a single day and a night'. It is not the destructive force of a localised tsunami but rather a tide, whose energy is spread-out all over the world. Buildings might be left intact as water filled-up around them and trees and plants survive unaffected; but animals and humans would be drowned by the sea unexpectedly rising up above their heads, perhaps on a sunny day with no warning of what was about to happen.

It is again important to note that this mechanism alone would not cause permanent submergence. Once the wobble is damped the shoreline should settle again very close to where it was before. To cause permanent sea-level change requires that the solid crust must deform. During a pole tide the crust would be subjected to abnormal stress, being alternately above and below its optimum geoid height. We should expect some parts of the crust to gradually sink or rise during this process all over the world. Again, we need not presume such a process to be destructive if it is spread over a period of years. We should expect earthquakes and perhaps volcanism at weak plate boundaries where these normally occur. However, the gradual adjustment that would be needed to align the figure axis to its new stability would be more likely, a 'folding', and increased lava flow at all the mid-ocean ridges. This is quite unlike the localised side-slip that we see along fault lines, or the sudden thrusts that cause uplift along subduction zones. It can only be a suggestion, but such 'folding' consequent upon pole shifts may be the mechanism which causes the sea-floor spreading that we observe in the geological record.

Ancient Fault Lines
On 1st November 1755 a strong earthquake followed by a tsunami struck the Portuguese capital of Lisbon. Estimated to have been of Richter magnitude 8.5 - 9.0 it is one of the largest earthquakes in recent history. The tsunami and fire that followed all but destroyed the old city and may have killed as many as 100,000 people in the region, as much of southern Algarve and Morocco was also affected. The effects of the

tsunami were felt as far away as Ireland; a surge of around eight feet was measured in Penzance Bay and the seismic waves caused a seich of two and a half feet in Loch Lomond. The wave may even have crossed the Atlantic to Brazil.

The epicentre of the 1755 quake has been much disputed, but modern investigations suggest that it was uplift along the transform fault that runs from the Azores to Gibraltar. More recent tremors in 2017 and 2018 awakened fears that tsunamis from this fault line are a major threat to Iberia and highlighted the need for a warning system.

Although north-west Europe is generally considered tectonically safe, there remain many inactive fault systems from ancient episodes of mountain building. The Great Glenn fault and the Highland Boundary fault in Scotland, together with the Dinorwic fault in the Irish Sea are the most obvious, and these faults extend out under the sea. These ancient features date from the coming-together of the continents that built the Caledonian and Appalachian mountains, long before the Atlantic Ocean opened-up some two-hundred million years ago. Other less ancient fault systems from the building of the Alps extend north through France and the Rhine valley and have also experienced quite large tremors during historical times. While these are not usually considered as active earthquake zones, we should still regard them as zones of weakness that could become active during an abnormal geological event. It has recently been proposed that in 1607 a sudden surge in the Bristol Channel may have been a tsunami, which originated along one of the old faults in the Celtic Sea.[12]

Anyone who would prefer a localised explanation for the Atlantis story and for the similar destruction described by Diodorus Siculus, might seek to rationalise it as just a memory of an ancient tsunami; perhaps occurring somewhere along the Atlantic coasts and similar to that of the Lisbon earthquake. Perhaps an earthquake and tsunami devastated a Neolithic settlement along the Iberian coast? It's a possibility – but beware that you don't fall into the same trap as those academics who would seek to localise Atlantis as Thera, or as a tsunami hitting ancient Crete. This requires you to selectively ignore part of the trail of evidence. The 1755 earthquake did not permanently sink any Atlantic islands; it did not trigger earthquakes in Greece, or cause the Nile Delta to expand. It has not been accompanied by a permanent climate shift in the Sahara and it did not cause Alpine glaciers to readvance. The scale of the causal event that could do all those things five-thousand years ago has to be a step-change up from a mere magnitude-9 earthquake.

Plato's Astronomy – a Neglected Source

Plato's *Timaeus* holds much more for us than just the Atlantis dialogues. In a later philosophical discussion he discusses the nature of the gods. He dare not question the origin of the Greek gods, their creation of the universe and their appearances as planets in the night sky; echoing here the later fears of Galileo to go against religious doctrine. He considers the stars and planets as living entities made of fire and capable of independent movement around the sky:

> ...the fixed stars...are living beings...always rotating in the same place and the same sense; the...planets we have already described. And the earth...winding as she does about the axis of the universe...[13]

With this description of the 'winding' or an additional rotation of the Earth he has puzzled many translators and commentators; for he has clearly already described the daily rotation of the stars and the apparent reverses of the planets; so what is this additional motion of the Earth? Some have deemed it to be the precession of the equinoxes, one of the long-term astronomical factors that affect the world's climate. The gravity of the planets pulls the oblique axis round in a complete cone over a period of 26,000 years. However most historians concur that the precession of the equinoxes was unknown to the early Greeks.

An alternative to consider is that Plato is here describing a wobble of the rotation axis. A similar motion is implied in Chinese astronomy, where the world is sometimes described as sliding up and down on the axis of the universe. Now any such wobble must be transient and ultimately would decay to nothing. As it is barely detectable today then it must represent a residual effect from a triggering event in the not-too-distant past. Was Plato describing a wobble that was still observable to naked eye astronomy in his own time (c.350 BC) or was he merely repeating older traditions? The latter explanation would be favoured, since later classical astronomers such as Eudoxus and Ptolemy do not mention it. If it recalls a wobble of the Earth's axis that had completely decayed by the classical era then it could offer another clue to when the causal event occurred. Climate scientists define the climatic period known as the *sub-boreal* (drier-cooler climate) that prevailed between about 3000 BC and 500 BC.[14] This might offer a hint to the date of the triggering event 5,000 years ago.

As a further comment at this point, one may note that Plato offers no discussion of the Greek calendar in his astronomical narrative. If he had devised an 11-year lunisolar cycle based on alternating five and six year periods, solely in order to mention it in a fictional account about Atlantis, then why does he not take the opportunity to further elaborate on it here? After all, it is potentially a far more accurate calendar than the 8-year cycle then used by the Greeks. This absence of explanation is a further clue that it is a fossil. The five and six year periods have come through all the way from the ultimate Egyptian source and Plato did not recognise what he had.

...and what can we learn from archaeology?
Archaeologists will tell us that farming and civilisation arose in the Middle East and only belatedly reached the west of Europe. This idea of diffusion is perhaps the last survival from the old archaeology that preceded the radiocarbon revolution, which has not yet been overturned. The accepted dates for the earliest agriculture in Spain and Gaul are around 5000 BC and for its arrival in Britain and Ireland somewhat later around 4400 BC. Archaeologists have long defined a transition from Mesolithic hunter-gatherer cultures to Neolithic farmers and pastoralists as shortly before 4000 BC, this new culture being deemed to have spread from the middle-east where Emmer wheat was first cultivated.

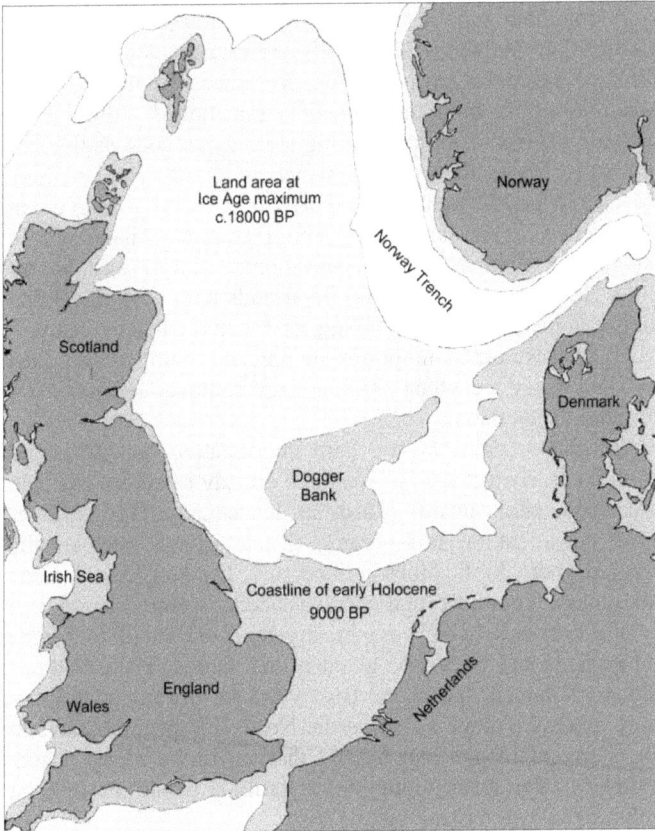

Figure 3.2 The Submergence of Doggerland

The standard glacio-eustatic model dictates that the eastern Irish Sea and the land-link between Britain and Ireland must have been submerged at around the same time that the North Sea land bridge to Europe was broken as the ice-sheets melted. This is not consistent with other physical evidence from around Irish Sea coasts; and the standard model would not give the ancient coastlines as they are described in the various legendary sources. These would require that the eastern Irish Sea remained above the sea as a dry plain long after the North Sea basin was flooded.

Throughout any discussion of the early archaeology of western Europe, you will typically see vague statements of the form: 'sea levels rose as the ice sheets melted...' and assumptions as to when societies transitioned from hunter gathering tribes to settled farming are only as good as the ice-age and climate theory upon which it all sits. We are here focused upon the reality (or not) of a legend preserved in Egypt, that a farming society with a great city, based on an island somewhere off European shores, existed well before 3000 BC. If so, then it would be at odds with everything that archaeologists have been telling us for more than a century. Many of these old assumptions are now also being undermined by DNA analyses, which promise a revolution as severe as that of radiocarbon dating.

To focus again: we are here investigating *how* an island could sink, not yet *where*. We have already ruled out regional volcanism and tsunami causes as inadequate. This removes from consideration mid-ocean volcanic islands such as the Azores where, in the nineteenth century, Donnelly could locate his sunken 'continent' and even subsidence–prone islands such as the Canaries, which are in any case too small to fit the dimensions and landscape given in our source. The coastlines around Morocco, Iberia and Biscay are steeply rising and it is only when we range as far as the North European continental shelf around Britain and Ireland that we find a region where extensive flat areas suitable for farming could have been submerged by the rising sea.

It is often said, that we know more about the surface of Mars than we know about the ocean floor – let alone what evidence might lay buried a few metres beneath it. Due to oil exploration, the North Sea is one region where we do know a little more than usual and the sub-surface topography has now been extensively mapped.

The realisation that Britain must formerly have been linked to continental Europe was a natural consequence of the Ice Age theory; as it follows that the shallow continental shelf must have been dry land when so much of the world's water was locked up in ice sheets. This assumption was further confirmed from the 1930s when fishing trawlers began to drag-up animal

bones and prehistoric stone arrowheads. The shallowest area in
the middle of the North Sea is around the Dogger Bank, so-
named after the Dutch boats that formerly fished the shallows
there. On navigation charts it may be seen that the rivers
Thames, Rhine and Elbe, formerly met in a broad river channel
running between the Dogger Bank and present day Denmark.
Hence this sunken region has taken-on the name of
'Doggerland'; though one may hope that one-day someone will
give it a more attractive name!

Conventional dating for the submergence of this low-lying
tundra region follows the glacio-eustatic theory which suggests
that it was drowned as the North American and European ice
caps gradually melted; and around 6,500 BC the land bridge
between England and the continent was finally broken.
However, Doggerland may have lingered above the sea for
much longer, perhaps as late as 5000 BC. The constraints of
glacio-eustatic theory therefore dictate that low-lying land
around western Britain and Ireland was similarly drowned at
this time. Soft sediment erosion around the east coast of
England continues to destroy any archaeological evidence of
ancient human occupation.

Confirmation that Northern Britain was separate from the
continent comes from the discovery in the early twenty-first
century of a mega-tsunami event around 6200 BC.[15] This was
caused by an underwater landslide off the coast of Norway and
is known as the *Storegga Slide*. At this time, a few parts of
ancient Doggerland may have persisted above the sea as low
islands and would have experienced the destructive effects of
this tsunami wave.

These landslides occurring within the deep underwater
canyon off the coast of Norway are believed to have been some
of the largest known undersea subsidence events and
worryingly, could occur again at any time. However, we should
note that their effects upon the coastline were not greatly
different to those observed after the recent Sumatra and
Japanese tsunami events; indeed that is how the specialists
recognise them.

It will not be surprising therefore, that Doggerland has from
time-to-time also been proposed as the site of Atlantis,

especially when it could be set close to the date given by Plato.[16] The Swedish author and scientist Ulf Erlingsson revived this theory in a quite independent way, linking it to Irish culture.[17] However, once again, the time-frame would not fit with that established here – the drowning of the North Sea is a few thousand years too old and Britain was already an island long before the earliest period of Dynastic Egypt. In fact it lies closer in time to that older flood event which caused the inundation of the Black Sea basin. Notwithstanding the tsunami event, the slow creep of sea level rise due to gradual ice-melt does not sound at all like the catastrophic event that legend requires.

In conclusion – just too many coincidences
We may see that the period around five thousand years ago (3200-2900BC), coinciding with predynastic Egypt, also corresponds to an era when the world climate took a turn for the worse. It is not easy to see how a mere regional earthquake or even a mega-tsunami could trigger the permanent climate transition that occurred world-wide around this time. We also see the correspondence with the short-term climate fluctuations recorded in tree-rings and ice-cores and fluctuations in Alpine glaciers. We may also note that it corresponds to unexplained changes in the carbon-14 environment; and with permanent sea level changes, uplift of land in central Asia and indications of submergence along the Atlantic coasts. How could Plato know about the pollen zones and the Hypsithermal; or that all these other things would correspond with 'the earliest period of the Egyptian state'; in order to place his supposed fiction at just the right era?

Perhaps it is safest not to stray too far beyond the pattern of evidence and just to conclude here that *an extraordinary geological event*, much more extensive than any major earthquake or regional tsunami, is required to explain all of the changes that occurred in the mid-Neolithic around 3000 BC. All one can really do is to present the pattern of evidence and say to the specialists: 'Here is the time and place that you should be looking'.

Notes and References

[1] As this book goes to press, a 31-kilometre-wide impact crater has been discovered beneath Hiawatha Glacier, northwest Greenland. The study loosely dates the impact between 12,000 and 3 million years old, but older than the formation of the ice-sheet. See: *Science Advances* 14 Nov 2018:Vol. 4, no. 11, eaar8173 DOI: 10.1126/sciadv.aar8173.

[2] I have not updated here the survey of sea level and climate evidence that I did in the 1980s & 1990s, which at that time required many published radiocarbon dates to be tree-ring adjusted. Field researchers tend to survey a single location over periods of time rather than looking for evidence across locations at a particular date. Of the hundreds of papers that I looked at only a small fraction held any unambiguous dating evidence.

[3] Dansgaard, W.; et al. (1989). *"The abrupt termination of the Younger Dryas climate event".* Nature. **339** (6225): 532–534.

[4] Reid, N, Nunn, Patrick, Sharpe, M. Foundation for Endangered Languages; 2014. Indigenous Australian stories and sea-level change.

[5] Principally this summary of sea level papers was contained in Chapter 6 of *The Atlantis Researches* or *Atlantis of the West*

[6] This is not strictly true. Apollodorus, for example, would place the Flood of Deucalion during the Bronze Age.

[7] These are the Blytt-Sernander pollen zones.

[8] Journal Proceedings of the National Academy of Sciences, May 2018.

[9] Herodotus, IV, 175-188

[10] Pliny, Natural History V,viii,45

[11] Dansgaard W. *Frozen Annals Greenland Ice Sheet Research.* Odder, Denmark: Narayana Press. p. 124. ISBN 87-990078-0-0.

[12] Haslett, Simon; Bryant, Edward (2004). *"The AD 1607 Coastal Flood in the Bristol Channel and Severn Estuary: Historical Records from Devon and Cornwall (UK)".* Archaeology in the Severn Estuary (15): 81–89.

[13] Plato, Timaeus, 8, 40-41

[14] In my earlier book *Under Ancient Skies* I attempted to trace evidence of such a motion by an examination of ancient eclipses. Unfortunately, the observations dates and timings in the ancient sources are too imprecise to offer conclusive proof of such a wobble.

[15]Bondevik, S; Lovholt, F; Harbitz, C; Stormo, S; Skjerdal, G (2006). "*The Storegga Slide Tsunami – Deposits, Run-up Heights and Radiocarbon Dating of the 8000-Year-Old Tsunami in the North Atlantic*". American Geophysical Union meeting.

[16] I shall not pursue here the literature concerning the Frisian *Oera Linda Book* that found its way into Nazi and new-age philosophies, or its revival by Robert Scrutton during the 1970s new-age fad.

[17] Erlingsson, U. (2004) *Atlantis from a Geographer's Perspective: Mapping the Fairy Land*. Lindorm Publishing. ISBN 0-9755946-0-5. p112

4

The Real Islands in the Atlantic

One may recall that wartime British film with vaudeville comedian Arthur Askey, called *The Ghost Train* about a steam-train that disappears into a tunnel and doesn't come out again. Says the stationmaster (in his Cornish-accented English): 'if it be a *real* train: where-do it come from; and where-do it go?' And so we must answer a similar question: where are the *real* large islands in the Atlantic Ocean that, five-thousand years ago, could have housed a civilisation such as Solon relates?

Having survived a chapter of exhausting science, the reader may perhaps be relieved to return to a less taxing discussion of ancient history and mythology. If we are to find the reality that underlies the story that the Egyptian priests related to Solon, then it is just as important to identify the geography. So far we have established the correct era as the late fourth millennium BC and that a destructive geological event, of greater power than anything recorded in historical times, is required to fit with all of the evidence from that era. The internal evidence of our sources tells us that a submergence event occurred along the Atlantic coast and not within the Mediterranean Sea; and that a civilization of ancient farmers was situated on a large island formerly accessible from the coast of Iberia.

The Unknown Atlantic Coast

An important factor in any examination of ancient history is to appreciate that the classical geographers of Solon and Plato's era knew almost nothing about the geography of northern and western Europe. There are two principal explanations for this; the first was the natural barrier presented by the Alps and the Carpathian Mountains with seemingly impenetrable forests beyond; the other was that the sea route beyond the Strait of Gibraltar was, throughout this period, blockaded by the

Phoenicians and later by the Carthaginians to protect their monopoly of the tin and silver trade.

For the era of Solon around 600 BC or even when we move forward to the time of Plato, we have recorded history for Greece and Egypt. The Egyptian Twenty-sixth Dynasty or the Greco-Persian wars are real history to us. Thanks to Herodotus we can even describe and name many of the tribes in North Africa and the Black Sea region – but for ancient Britain, for Gaul and Germany we have scarcely any idea of the tribes; or even the name of one single king from this era; and we are still only half-way back to 3000 BC. Yes, we do have the 'magical' Celtic legends for Ireland and Britain and we shall look further at some of these in the next chapter; but they are quite timeless and can be at best regarded as pseudo-history.

From about 900 BC the Phoenicians of Tyre began to expand their maritime trading westwards, in search of new sources of revenue to pay their tributes to the Assyrian Empire. Around 814 BC they founded the colony at Carthage, which by the era of Solon had become an independent city. Initially it was a half-way point on the voyage to Gades, the modern Spanish city of Cadiz; and the surrounding region of Tartessos from which they obtained their silver. From Gades they could range further north into the Atlantic Ocean and monopolise the sources of tin. Ships from Greece and other maritime trading nations, which during the Bronze Age may have passed freely into the Atlantic, could no longer get through. We may presume that trade in the reverse direction was also barred. Gradually, knowledge of what lay beyond the Straits began to fade and pass into legend, everywhere that is, except in Carthage.

A quotation from Herodotus illustrates the void that existed in Greek knowledge of the west. He first tells us first about the Phoenicians' circumnavigation of Africa and of the Persian sea voyages to India:

> With Europe however, the case is different; for no-one has ever determined whether or not there is a sea either to the east or to the north of it; all we know is that in length it is equal to Asia and Libya combined.[1]

He knows of the River Danube, which he says rises among the Celts (i.e. southern Germany) but knows nothing of the Alps or the Carpathians; he does however describe a river Alpis and a river Carpis, flowing northward. Beyond the Celts, he mentions the *Cynetes* or *Cynesians* as the furthest west of all nations; and in another place that both the Celts and Cynetes lived beyond the Straits of Gibraltar – but he does not say that they lived near the sea. Some commentators have located the Cynetes in Spain, or perhaps the Cantii of Britain. As for Iberia, Herodotus makes scant mention, save that he knew of the existence of Tartessos, beyond the Straits.

Furthermore, Herodotus refused to believe second-hand reports that we know to be true. He dismissed the very existence of the Cassiterite Islands whence the tin originated. Neither could he believe in the existence of a sea in that direction, from which came amber and gold via the River Eridanus.

> ...I have never found anyone who could give me first-hand information of the existence of a sea beyond Europe to the north and west. Yet it cannot be disputed that tin and amber do come to us from what one might call the ends of the earth.

Amber, we may be certain, came from the eastern shores of the Baltic Sea – but Herodotus knew of no-one who had seen these places.[2]

Some time in his youth, before his three-month sojourn in Egypt, Herodotus made an expedition to the Black Sea. There he made enquiries about the rivers and geography to the north and about the Persian campaign against the Scythians. He describes the ethnography of the region as an aside to his main subject – the Greco-Persian Wars. The great rivers of southern Russia were another route by which trade goods from northern Europe could reach the Mediterranean. Around 500 BC the nomadic Scythians ranged over much of what we now know as southern Russia and Ukraine, from which some centuries earlier they had displaced its former occupants, the Cimmerians; somewhere beyond them were Issedones and Arimaspians.

Homer and Pindar had composed poetry about *Hyperboreans* and Herodotus expected that these people existed somewhere in the north; and so he made enquiries of his hosts, but no-one he asked seemed to know anything about them. He cites a Greek poem called the *Arimaspia* by Aristeas, a poet of Marmora island who wrote sometime before Homer – perhaps 800-1000 BC – but we cannot be more precise; so this is the era that the ethnography is describing. Aristeas journeyed only as far as the land of the Issedones. We can't be certain where this was either, but Ptolemy would later describe Essedones along the Baltic coast (the Finnic Estonians). The Issedones told Aristeas about the Hyperboreans whose country lay somewhere beyond the Arimaspians and griffins and whose land, Herodotus says, came down to the sea – this despite having said earlier that no-one knew whether there was a sea in that direction. His vague geography has had commentators heading off into central Asia to find the Hyperboreans. Fortunately there are other sources to clarify this.

Another reference to the Hyperboreans comes via Hecataeus of Abdera, a later author who clearly did know that there was a sea to the west of Europe.[3] His description is discussed further below, and would unmistakably place the Hyperboreans in the north-west beyond the territory of the Celts. However, the name itself is Greek meaning: 'people beyond the North'; we don't therefore know by what name they called themselves – unless they be one and the same with the Cynetes.

It is likely that very few Greeks of the classical era really understood what their mythology was all about, any more than we know today. We may study mythographers such as Apollodorus, who wrote in the first century BC, yet emerge not a lot wiser and with even more jumbled myths that still cannot rationalise to history. Every Greek might learn the story of Heracles; or accept that the Isles of the Blessed, where lay the plain of Elysium and the Garden of the Hesperides, might exist in some vague location to the west. They need enquire no further about where the gods were; any more than a catholic in Bolivia doubts the existence, in ancient times, of the holy-land in far away Israel.

Some legends, such as those in Hesiod and Homer, do perhaps recall events of the Bronze Age and earlier, before the Greek 'dark age' around 1200 BC drew a veil over our knowledge. In these poetic sources we find many tantalising references to the Hyperboreans, which confirm their ancient contacts with the Greeks. However, it is important to note that these poems were never intended to record history; rather they were a form of popular entertainment to be sung or performed aloud. It's rather like trying to reconstruct a history of America, when all you have is a song about Davy Crockett who was a king of the wild frontier! Real historical people may be set in real historical places. The content may be fiction but the people and settings give us clues to real ancient geography as useful as any artefact dug from the soil.

In Homer's *Odyssey*, the hero Odysseus loses his way while returning home from the Trojan War and embarks upon a sea voyage to the west, where he is detained for seven years by the goddess Calypso on the misty island of Ogygia. Along the way he visits the land of the dead and, among other places, the land of the cannibal Laestrygones. We need not dwell too long on the geography. Classical scholars have long argued whether Homer describes real or imaginary places, but successful fiction has to be set in locations that the audience will recognise.

Commentators since at least the time of Polybius and Strabo have considered the Odyssey to record a mere fictional voyage around Italy and the Mediterranean; yet many anomalies suggest that the setting can only be in the Atlantic. For example, we may note the description of the land of the Laestrygonians, where morning follows nightfall by only a short delay; and where a sheepherder who can do without sleep might earn two wages in a day. Such a description of the midnight sun phenomenon could only apply at latitudes near the Arctic Circle, perhaps Orkney or Shetland. The writer who understood this most convincingly was Plutarch, who in Roman times informs us that much of Greek mythology concurred with native British traditions.

Diodorus and the Islands of the West

In the histories of Diodorus Siculus we find two quite different accounts of islands in the west; three if we include his contemporary knowledge of Caesar's British campaigns. His own era was 80-20 BC and his history ran from the earliest times up to 60 BC. He provides a description of Britain as a triangular island larger than his homeland of Sicily; and his details about the tin mining in Cornwall can only have come ultimately from Pytheas of Marseille.[4]

Diodorus had quite a different perspective from other Greek authors, as many of his older sources were drawn ultimately from North Africa and Egypt via Dionysius and the Kyklos encyclopaedia; and although he wrote in Greek he may never actually have visited Athens himself. We would expect more authentic knowledge of the Atlantic Ocean to have been recorded by sailors of Carthage who actually traded there; and for these sources to have survived in the Library at Alexandria long after the Romans had reduced Carthage to rubble.

Another story tells us a little more about the Hyperboreans and it will be worthwhile here to summarise and quote parts of it. C. H. Oldfather and many other commentators have considered this to be a description of Britain, and that its temples remember the Neolithic stone circles.[5]

> Of those who have written about the ancient myths, Hecataeus and certain others that in the regions beyond the land of the Celts there lies in the ocean an island no smaller than Sicily. This island the account continues is situated in the north and is inhabited by the Hyperboreans...[6]

He describes the mild climate of the island and that it could produce two crops each year; and the magnificent 'spherical' temple of Apollo, which due to the calendrical references of Hecataeus has been likened to Stonehenge, or perhaps one of the other astronomically aligned stone circles.[7] The Boreadae who supervised the temple also cherished an ancient relationship to the Greeks.

> The Hyperboreans also have a language, we are informed, which is peculiar to them, and are most friendly disposed

towards the Greeks and especially towards the Athenians and the Delians, who have inherited this good-will from most ancient times.

Now what could he mean here by: 'a language peculiar to them'? Are not all foreign languages peculiar? He means that it is unlike any language that he knows; perhaps a clue that it was non-Indo-European and therefore did not follow the grammar rules of Greek. He does not, for example, remark that the language of the Celts was peculiar. This is another of those useful fossil-clues for us to investigate. We may conclude from this that at the Hyperborean era, the inhabitants of the island beyond the Celts were *not yet* speaking a Celtic language.

A third description is also given of an unknown island in the Atlantic, perhaps recording a genuine discovery by some explorer of ancient times, or more likely a diversionary tale perpetuated by the Phoenicians:

But now that we have discussed...the islands within the Pillars of Hercules, we shall give an account of those which are in the ocean. For there lies out in the deep off Libya an island of considerable size, and...distant from Libya a voyage of a number of days to the west. Its land is fruitful, much of it being mountainous and not a little being a level plain of surpassing beauty. Through it flow navigable rivers which are used for irrigation, and the island contains many parks planted with trees of every variety...which are traversed by streams of sweet water; on it also are private villas of costly construction...

It should be immediately apparent here that there is a similarity between this paradise-island and the mountainous island with its irrigated plain and rivers, as described to Solon; and with those similar recollections of the Elysian Fields in the west that we find in the Egyptian Book of the Dead papyri. Note the fossil again: the plain is described as a 'level plain of surpassing beauty'. Not just any ordinary plain; clearly this one was special. It cannot be describing small islands such as the Azores or Canary Islands, for these do not have plains or navigable rivers.

It is sometimes necessary here to use quotations to demonstrate that these correspondences *are present in the ancient sources* and are not some contrivance of the present author. Comparisons are only made possible due to the independent translations from Greek or Egyptian by specialist scholars who hold no such questionable agenda. The molecules of a legend, the 'fossils', are indivisible. They survive retelling and translation and may have done so for thousands of years.

Although C. H. Oldfather (whose translations are followed here) refused to consider where this magical island might lie, we may be certain that it is another view of Britain or Ireland, in some much earlier era, even older than the Hyperboreans. Either it describes the Emerald Isle or England's green and pleasant land. We can forgive Diodorus for being unsure whether some other large island existed out in the broad Atlantic and for failing to rationalise that he had three views of the same place from different eras and sources; but armed with our twenty-first century knowledge we have no such excuse. The idyllic view of a lush green island and its fertile plain can only have been a very ancient vision of Britain or Ireland recorded by someone from a dry desert land who would seldom experience rain falling from the sky; and to whom gently cooling breezes were a refreshing phenomenon. We may even note the climate marker in all these myths, which would assign it to that warm period of the Atlantic pollen zone and thus place its origins earlier than 3000 BC.

> And, speaking generally, the climate of the island is so altogether mild that it produces in abundance the fruits of trees and other seasonal fruits for the larger part of the year, so that it would appear that the island, because of its exceptional felicity, were a dwelling place of a race of gods and not of men.

One cannot help but see here the same place that Diodorus described earlier as the home of the gods, whence the Libyan Amazons claimed their ancestors had come; and Diodorus is inviting us to see this similarity even though he does not know where the island actually lay, or whether it be real.

We encounter this view of an island paradise yet again in Plutarch's *Life of Sertorius*; that rebellious Roman general, who between 80 and 72 BC fought a guerrilla war with Rome and would prove impossible to dislodge from his base in Spain – until his own allies turned on him. While visiting Gades, Sertorius met with some sailors who had just returned from 'the Atlantic Islands' out in the Ocean opposite Africa; and momentarily Sertorius wished that he too could just sail away and abandon his worries. Plutarch describes two islands separated by a narrow strait, which he likens to The Islands of the Blessed. He comments on their amiable climate:

> ...while the south and west winds that envelope the islands from the sea sometimes bring in their train soft and intermittent showers and gently nourish the soil. Therefore a firm belief has made its way, even to the Barbarians, that here is the Elysian Field and the abode of the blessed, of which Homer sang.[8]

The real islands that the sailors had visited were most likely the two islands of Madeira, to which Plutarch has (here) interpolated features of those imaginary lush green islands of which he has heard, in the usual way that happens with the retelling of legends. Even by early Roman times, certainty about what lay out in the Atlantic Ocean had progressed only a little further than in the era of Solon – and indeed would not do so until Columbus.

The confusion of this paradise island with the source of tin and silver is further confirmed by Diodorus Siculus. Once again a quotation is necessary to demonstrate how it fits with the earlier cited evidence:

> In ancient times this island remained undiscovered because of its distance from the entire inhabited world, but it was discovered at a later period for the following reason. The Phoenicians, who from ancient times on made voyages continually for purposes of trade, planted colonies throughout Libya and not a few as well in the western parts of Europe...they founded a city on the shores of Europe, and since the land formed a peninsula they called the city Gadeira.

...The Phoenicians then, while exploring the coast outside the Pillars for the reasons we have stated and while sailing along the shore of Libya, were driven by strong winds a great distance out into the ocean. And after being storm-tossed for many days were carried ashore on the island we mentioned above and when they observed its felicity and nature they caused it to be known to all men.[9]

We may see that the belief that this island was a real place dates from the Phoenician era and was available well before 600 BC.

Figure 4.1 Islands 'opposite' Iberia.
The only real large islands accessible from Iberia are Britain and Ireland, but these lie 'opposite' the north coast rather than Gadeira. The Ancient routes of the tin trade followed the coast of Spain and Gaul.

Diodorus also tells us that later, the Tyrrhenians (Etruscans) had wanted to send a colony to the fabled Atlantic Island but that the Carthaginians refused permission, ostensibly because they feared that their own people would want to migrate there. Of course, we can be certain that they could not allow the voyage because the paradise island did not really exist; and they could not direct them to the real islands further north lest they lose their monopoly of the tin. The Etruscans had originally possessed their own local source of tin in Tuscany, but around this time it became exhausted. This saga tells us, if we did not know already, that the Carthaginians continued to circulate this ancient fable in order to hide where the tin and silver really came from.

In Search of Tin and Bronze
Bronze is an alloy of copper with around 12% of the element tin, rendering it harder than pure copper; although ancient bronze often had other impurities and contained much less tin due to the difficulty of obtaining it. From where the people of the Bronze Age obtained their tin; and indeed, where they were first mixed together; is a question that has long vexed archaeologists and historians. Copper in the ancient near-east came predominantly from Cyprus, which is of course what the name of the island means. It provided an economical source of the metal in Egypt from predynastic times.

The transition to the so-called Bronze Age is put as early as the proto-dynastic era for Egypt, but is not generally believed to have reached the west of Europe until much later, around 2500 BC. However, earlier bronze artefacts from perhaps 5000 BC come from Iran and China; and also perhaps around 4500 BC from the Vinča culture in modern Serbia. This is not to say that bronze weapons had completely displaced copper and stone weaponry, as this would have required a regular and economical source of tin. Copper tools continued to be used along with stone tools, for example during the Pyramid age – the copper or 'Chalcolithic' age is really only a phenomenon for near eastern archaeology. In the west of Europe the transition from flint tools to bronze seems to have happened without an intermediate stage.

Although Herodotus gives us our earliest mention of the Cassiterite Islands in classical sources, he did not actually know where they were. Cassiterite mineral is almost pure tin oxide; some early bronze and brass artefacts contained other impurities, such as arsenic, which gives off a poisonous vapour when smelted; hence the early metalworkers placed a value on the pure cassiterite ore. The principal sources in Europe are: Cornwall, the estuary of the Loire and also Lusitania (modern Portugal). Evidence of the ancient tin mines is of course lost as they have been destroyed by later mining – but we should not rule-out that many early mines have been drowned by the rising sea, as many of the seams extend far below sea level. Recent archaeological discoveries would suggest that almost all early bronze artefacts from Europe bear the chemical composition of Cornish tin.[10]

This question of where tin and copper were obtained has much relevance for the present discussion because it is another of those 'mythological fossils' in Plato's Atlantis story that merits separate examination. He mentions copper or brass in various contexts and that tin was available in sufficient quantity to make a brazen display on the walls of the city. Therefore, if it were a real place then there must have been a convenient source of these metals nearby. Recall the *Critias*:

> The entire circuit of the wall, which went round the outermost zone, they covered with a coating of brass, and the circuit of the next wall they coated with tin, and the third, which encompassed the citadel, flashed with the red light of orichalcum.

It is possible that awareness of such metals existed in the west during the fourth millennium BC but that, as Plato relates, such knowledge was lost in a dark-age that severed contact with the Atlantic coast for many centuries.

Strabo and the Voyage of Pytheas

Move forward now to early Roman times. It was not really until Rome expanded into Gaul in the mid-first century BC that we gain any detailed knowledge from our sources about the tribes and geography of the Atlantic coast.

The inhabitants of Atlantic Europe: the Celts and their neighbours must have known from the earliest times where the tin and amber came from. They didn't need any Phoenicians to discover it for them, nor to understand why they wanted it. They were quite capable of trading their tin themselves into Mediterranean markets – until these more advanced and warlike nations came in ships to colonise their ports and mines. From 600 BC onward, tin was among the principal trades of the Greek colony at Marseilles, who transported it overland from Cornwall and the Loire estuary via the rivers Rhone and Garonne.

Around the year 285 BC the navigator Pytheas set-out from Marseille to explore the sea-route that the Carthaginians were using to reach the fabled tin islands; he successfully evaded their blockade at the straits and headed into the Atlantic Ocean. It is unfortunate that his own account of the voyage, called *On the Ocean,* does not survive, but fragments of his itinerary are quoted by other geographers; especially the Roman writers Polybius and Strabo; and also by Posidonius whose 52-volume history is another sad loss. These conservative-minded authors simply refused to believe Pytheas' account of places that we may recognise today and we need have no doubt that the voyage was real. His measurements of latitude at various points were surprisingly accurate and were used by later astronomers, such as Geminus and Eratosthenes to estimate the size of the Earth. From the unsympathetic accounts of his later commentators we may work out approximately the places that Pytheas visited.

We may see that the voyage took Pytheas north along the coast of Iberia where he turned east at Cape Finisterre and followed the northern coast of Spain, until he reached the port of Corbilo at the mouth of the Loire, Here he noted that tin from Britain was unshipped for the overland journey to Marseille. He then followed the route of the Carthaginians to Uxisama (Ushant) where he notes a tribe called the Osismii. Next, he set course north, landing after a day near Belerium (Land's End). He conversed with the inhabitants about the source of the tin ore and learned how it was shipped south along the route that he had travelled.

The route further north is less certain, but we may be sure that Pytheas was following a well-known itinerary confirmed by the local inhabitants. We may see that although his primary motives (and of those who funded his voyage) may have been commercial intelligence, he would also record his journey in a scientific manner as good as any later Renaissance navigator. Our various sources record that he circumnavigated Britain and reached the northern cape of Orca (Dunnet Head). He may have put-in more than once on the east coast of Ireland – for how else could the descriptions of its inhabitants have reached Strabo? He may have also visited, or was told about, another large island called *Thule*, which supposedly lay six days sail north from Britain.

Many have commented on where the island of Thule actually was but the fragmentary record is unhelpful. We cannot determine whether the geography of Pytheas was erroneous or whether his later commentators have misinterpreted him. We do not know whether he saw this place himself or was reporting local information. It is unfortunate that Strabo, our principal source, did not believe it existed nor the 'frozen sea' around it; that it was close to the Arctic Circle is clear from the description of its short summer nights and the latitude calculated by Geminus. Probably however, the confusion is easily explained. On his voyage north, Pytheas must have passed between the Scottish Mainland and the Outer Hebrides, and noting the length of its coast assumed it to extend equally far to the north. Thule may therefore be a conflation of the Isle of Lewis, with some features of Iceland or Norway added from other reports.

We are told that Pytheas then returned south via the 'Amber Coast'; which suggests that he crossed the North Sea and followed the coastline down to the Rhine, or that he followed the east coast of Britain before crossing to the Rhine. He may just have been reporting local information again. We have no reason to believe that he ventured into the Baltic Sea to the ultimate source of amber, but it gives us another clue to the commercial motives of his journey. In Roman times, amber was traded to the Mediterranean overland across central Europe and earlier via the Black Sea route.

Of the Cornish sources of tin we again have the description of Diodorus Siculus, which has without doubt come principally from Pytheas. He describes Britain as roughly triangular in shape and gives the dimensions between its three capes. He mentions Land's End and gives its distance as four days from the mainland.

> The inhabitants of Britain who dwell about the promontory known as Belerium are especially hospitable to strangers and have adopted a civilised manner...they it is who work the tin, treating the bed which bears it in an ingenious manner...On the island of Ictis [St Michael's Mount] the merchants purchase the tin of the natives and carry it from there across the strait to Galatia or Gaul...[11]

And of Strabo who again describes Britain as triangular:

> ...and its longest side stretches parallel to Celtica...the Celtic length that extends from the outlets of the Rhenus [Rhine] as far as those northern ends of the Pyrenees that are near Aquitania...[12]

There are three geographical errors here; the south coast is not the longest side, but the shortest, as other contemporary writers well knew. He seems to know only of the north coast of Cornwall and Devon and extrapolated it to make a smaller triangle; this would fit with his belief that Ireland lay to the north of Britain. Also, Strabo did not know of the existence of the Brittany peninsula, for in discussing Pytheas, he says:

> ...from Cantium...as far as that westerly end of the island [Land's End] which lies opposite the Aquitanian Pyrenees.

The reason why he has the impression that Land's End was closer to Spain becomes clear when he discusses the four routes that were used by seafarers to reach the island:

> There are four passages which are habitually used in crossing to the island, those which begin at the mouths of the rivers...The Rhenus [Rhine], the Sequana [Seine], the Liger [Loire], and the Garumna [Garonne]...

Because he believes that the coastline forms an almost-straight line between the mouths of the Garonne and the Rhine, he has not realised that the passage to Britain runs along the coast as far as the cape of Brittany before it must turn north to risk the open sea. This tells us indirectly that mariners from Spain and the Pyrenean coast would all hug the shore (as did Pytheas) rather than risk the notoriously stormy crossing of the Bay of Biscay. To reach Ireland would be a continuation of this route. Ireland, he calls *Ierne* and describes it as a large rectangular island 'which stretches parallel to Britain on the north'. Beyond that, following Pytheas again, he reluctantly places *Thule* at the limits of the known world.

The errors in Strabo's geography are nowhere more apparent than in his description of the Brittany peninsula:

> Secondly there are the Osismii (whom Pytheas calls the Ostimii), who live on a promontory that projects quite far out into the ocean, though not so far as he and those who have trusted him say.

Since Strabo refused to believe the information from Pytheas then he must have been falling back upon older descriptions which led him to construct the curious geography for the coast of Spain and France. He therefore quite unintentionally preserves for us that earlier geography. These older sources appear to have placed the source of the tin on the Cassiterite Islands lying 'opposite' the coast of Iberia, as we may see again from Diodorus Siculus:

> For there are many mines of tin in the country above Lusitania and the islets which lie off Iberia out in the ocean and are called because of that fact the Cassiterides.[13]

We may therefore see perhaps, why we have so many accounts of islands lying 'opposite' Gades and out in the open ocean.

The geography that we find in *Critias* also mentions a peninsula lying opposite Gades which was assigned to one of the ten kings. If we look on a map for a real large island with a peninsula lying 'opposite' to Iberia then there is only one candidate and that would be the peninsula of Devon and

Cornwall. As the source of tin it must always have held political importance from the earliest discovery of Bronze.

This older Atlantic geography that we find fossilised in Strabo must have come via the Phoenicians and Carthaginians, and was full of inaccuracies, which they either deliberately put about or at least did not refute. Anyone who defeated their blockade would head west into the open ocean in search of the fabled Cassiterite Islands and never be seen again! At best, if they dared continue against the winds and currents for several days, they might encounter the uninhabited Azores.

When Caesar's conquest reached the Atlantic coast of Gaul in 57 BC he would describe the tribes of the Brittany peninsula: the Osimii, with the Veneti and Pictones, prosperous seafaring nations around the mouth of the Loire. He and his general Publius Crassus fought a great sea battle with the Veneti and destroyed most of their fleet. From that era the Romans would take over the tin production on the Cassiterite sources in the Loire; and it is from after this date that most on our surviving descriptions of the trade were recorded.

Procopius and the Land of the Dead

The casual remark by Plutarch quoted above: that knowledge of the Elysian Fields and the abode of the blessed dead had reached even to the Barbarian nations is confirmed by a perhaps surprising source – the sixth century Byzantine historian Procopius of Caesarea.

Now firstly, we should note the context. Procopius is giving us an interesting aside to his central theme, which was an account of the wars between the late Roman Empire and the Barbarian invaders. As part of his history of the Gothic war around AD 550 he is describing the post-Roman settlement of the Anglo-Saxons in Britain. His poor knowledge of the contemporary geography perhaps reflects his own remoteness; but he is one of only a few sources we possess for the dark-age that followed the collapse of Roman Britain. Somewhere in this dark period historians believe that a King Arthur is hiding from the archaeologists! Procopius was actually 'pulling on an interesting thread', a digression that had little to do with his primary subject – and we are fortunate that he did. So often we have been left with these fragments of ancient history when the original source has not survived; and it is sad that historians who should know better have questioned the value of his Anglo-Saxon references – simply because he recorded for us some fascinating legends.

The story concerns an island called *Brittia*, which Procopius believed lay north of the former Roman province of Britannia. His translator, H.B. Dewing, considered it to be Denmark, but it is clearly describing the northern part of England opposite the Rhine. He names the tribes there as Angles, Frisians and Brittones; these latter were, at that time (around AD 550) probably the independent Welsh-speaking kingdoms of Elmet, and Reged. Even at this comparatively recent date Procopius was still considering his Brittia and the former Roman province of Britannia to be two separate islands facing Spain, with Thule beyond.

Procopius offers us another memory of the agricultural plain and its mild climate that sounds much like those encountered previously in the various descriptions of the paradise island out in mid-ocean. The men of ancient Brittia, he says, built a long wall to separate the habitable land from the less favoured region to the west, where men and animals would immediately fall dead should they venture:

> For to the east of the wall there is a salubrious air, changing with the seasons, being moderately warm in summer and cool in winter. And many people dwell there…and the trees abound with fruits which ripen at the fitting season, and the corn lands flourish as abundantly as any…

Procopius tells how the souls of the dead were ferried to this place (Brittia), which he has previously described. Once again, only a quotation can suffice to illustrate.

> …They say, then, that the souls of men who die are always conveyed to this place…I shall presently explain, having many a time heard the people there most earnestly describe it…Along the coast of the ocean which lies opposite the island of Brittia there are numerous villages. These are inhabited by men who fish with nets…or carry on a sea-trade with this

island, being in other respects subject to the Franks, but never making them any payment of tribute, that burden having been remitted to them from ancient times on account...of a certain service, which will here be described.

Now, this refers to the coast of Brittany around the period when Britons, mainly from Cornwall, were emigrating there to escape the Saxons; to the same regions where half a millennium earlier Julius Caesar had conquered the sea-faring *Veneti*, so we may be confident that it was originally they who had performed this ancient service of ferrying the souls of the dead to the west of Britain.

Of course, we may be sure that this was just a symbolic funeral rite, as Procopius describes how the boats would put to sea laden down by the weight of all the souls, and would reach Brittia after just an hour, where their invisible passengers would disembark. The fishermen would then hurry home unburdened, just as a voice from the island could be heard announcing the names of the new arrivals. The significant detail in all this is that the land of the dead is no longer being described as the abode of the blessed in some vague western location; but is this time associated with an identifiable place: the west of Britain; and is given by people who already reside along the Atlantic coast.

Procopius could only think that the entire tale had the properties of a dream about it; and indeed we may see that it has many parallels with other timeless Celtic legends. He says that he published the story only because: 'it was constantly being published by countless persons' and so that he 'should not gain a lasting reputation for ignorance of what takes place there'.

Such and so Great an Island

And so to return to that other Byzantine author Proclus whose commentary on Plato's Atlantis was introduced in Chapter 2. We may not know much about the author *Marcellus* whom Proclus cites, or precisely at what era he wrote, but he is clearly citing more than one earlier author; he thus demonstrates that written history and geography about the Atlantic coast of

Europe did exist in classical times – its is just that our surviving sources have not preserved them. Perhaps one of these historians was Pytheas, or Hecataeus who wrote about the Hyperboreans, but the core subject of Marcellus was African history, so his sources were probably also African: Carthaginian or Egyptian. There must have been numerous books and scrolls whose names and authors we can never know.

> That such and so great an island existed, is evinced by those who have composed histories of things relative to the external sea.

The authors (plural) tell of islands out in the sea. These are not myths and legends; these were real places, existing in their own times.

> For they relate that in their times there were seven islands in the Atlantic sea, sacred to Proserpine [Persephone]: and besides these three others of immense magnitude; one of which was sacred to Pluto [Hades], another to Ammon [Zeus], and another which is in the middle of these, and is of a thousand stadia, to Neptune [Poseidon].

The seven islands may perhaps record the Canary Islands, or maybe the Isles of Scilly. We are not told that these were large islands. However, the three islands of 'immense magnitude' can only refer to Britain and Ireland; as these are the only real islands that would fit this description; unless perhaps you wish to consider Iceland or Scandinavia.

> And besides this, that the inhabitants of this last island preserved the memory of the prodigious magnitude of the Atlantic Island, as related by their ancestors; and of its governing for many periods all the islands in the Atlantic sea. And such is the relation of Marcellus in his Aethiopic history.

This legend was kept alive by the people of this central island in the time of an unknown classical writer. The author is telling us that inhabitants of this island, which was sacred to Poseidon, preserved a history or legend of an earlier time when it had been the centre of a large maritime empire – which we can assert (since it is mentioned in the context of African history) extended as far south as the African coast and the nearby islands. If you would look for a large island at the centre of Britain and Ireland, there are only two candidates on the map: either it was Anglesey or it was the Isle of Man.

So consider something that really should be obvious. If Plato's account of a civilisation on an Atlantic island should turn-out to be mere fiction after all then we can all forget about it and go home! However, if it were a real place then ask-yourself: which are the *real* large islands off the Atlantic coast of Europe where it could possibly have been? You have Britain, or you have or Ireland, or you have both. There are no other candidates. Either it was there, or it was nowhere! It follows logically, since these islands still exist today, that the entire island as described by Plato was not submerged, but only the low lying parts of it. The city, the fields, and the other features that he describes, *if they be real*, must therefore lie submerged off the coast around these islands. Forget all the fantasies that you have heard. Read the above summary again and satisfy yourself that there is no flaw in the logic.

And so to America?

Every few years, it would seem some publisher or television producer will come out with a new conjecture about Atlantis, linking it with America or the Caribbean. So perhaps at this point it is pertinent to see how that sits with the chain of evidence.

Plato's *Timaeus* holds that remark, which implies that before the catastrophe in the west, the inhabitants of the Atlantic Island were regularly crossing the Atlantic Ocean.

> ... [it] was the way to other islands, and from these you might pass to the whole of the opposite continent which surrounded the true ocean.

This knowledge of navigation was then lost in the dark-age that followed and sea travel in the Atlantic was curtailed. So how realistic is it to propose that, before 3000 BC (in this context) ancient Europeans knew of the existence of America?

Firstly, we should note that nothing here implies a regular commerce and might record no more than a single discovery by an ancient explorer. When Portuguese explorers discovered the Azores in 1432 they were uninhabited and no commercial reason existed to go there. With Madeira and the Canary Islands the case is stronger. From Columbus onwards all transatlantic sailors would have to go this far south to catch the Trade Winds that would carry them across to the Caribbean; but the voyage of Columbus took fully five weeks at sea and the same to return. There is nothing in the trail of evidence, which has brought us to this point, to suggest that such long voyages away from the coast were attempted in ancient times; still less does it record any visitors from the other direction.

Perhaps the true memory of these ancient voyages is best preserved in the writings of Plutarch (AD 46-125). He travelled widely within the Roman Empire and certainly held a different view of the relationship between Greek and Celtic myths than we find elsewhere. In his *Obsolescence of Oracles* he tells us about Demetrius of Tarsus, an explorer sent out from Rome some time before AD 83 to report on the religion of the Druids. To put this into context, it was around this time in AD 78 that Agricola would all but exterminate the Druids in Anglesey and embark on his campaign in Caledonia. It is therefore likely that Demetrius travelled with him for a part of that expedition; and it is almost certainly from him that Plutarch gained his opinion that Greek myths of the Isles of the Blessed were well known among the Barbarians.

All that we know of Demetrius is what Plutarch tells us; and some of his descriptions of British myths can only have come from someone who had actually visited the island. Two votive offerings written in Greek, now in the Yorkshire Museum, are believed to have been left by this Demetrius.[14] However, to follow all the interesting threads in Plutarch's *Moralia* would again lead us too far off-subject; suffice to say that Demetrius was a guest of Plutarch on his way home to Tarsus and they discussed his experiences. He says that, on the orders of the emperor (presumably Vespasian) he had made a voyage to one of the nearer islands off the British coast that was inhabited only by a group of holy men held inviolate by the Britons. They told him about another nearby island 'where Cronus sleeps' guarded by demigods and attendants.

In another essay: *The Face on the Moon* Plutarch elaborates further on this island, which he equates with Homer's Ogygia. He says that a voyage of five days north-west of Britain, towards the summer sunset, lay this island 'and three others equally distant from it and from one-another'. On one of these, he says, lay the sleeping Cronus. Five days voyage from where? We do not know the starting point. We may conjecture but unfortunately we cannot pinpoint Ogygia with certainty. To quote Plutarch here:

> The great mainland, by which the great ocean is encircled, while not so far from the other islands, is about five thousand stades from Ogygia, the voyage being made by oar...

He mentions here the congealed sea, which is perhaps a misunderstanding of the ice-flows mentioned by Pytheas. When later Plutarch again mentions 'the mainland' It is not clear whether he is referring to the 'great mainland', or whether he intends the mainland of northern Britain; for he says that Greeks dwelt there and on the outlying islands, having migrated there in the time of Heracles. He gives the latitude of this settlement as about the same as the Caspian Sea. Can we really believe that there were identifiable Greek settlers living on the coast of Scotland in early Roman times? For that is what Plutarch implies. One day, perhaps, modern DNA techniques will answer this question.

Many commentators have supposed Plutarch's account to be entirely based on Plato's Atlantis and the 'Thule' of Pytheas. However, Demetrius was surely a real person. It is the astronomy and the scheduling of these voyages that confirm their authenticity. Plutarch relates that the voyages were made each thirty years when the planet Saturn, 'the star of Cronus', returned to the constellation Taurus; and we know from other sources that the period of thirty years was very important in the Druids' calendar.[15]

The essay is actually following the story of 'the stranger' (whose identity is lost along with the beginning of the source manuscript). He was a holy man who, having completed his thirty years on the distant island, wished to return to the mainland and visit Carthage because Cronus was so highly revered there; and this was where he met with Sextius Sulla – the Carthaginian who is relating the story to Plutarch. We may therefore see that the motivation for the Atlantic voyages was entirely religious and were ongoing right up to Roman times; for on this island was an oracle, where the faithful would receive the prophecies in the dreams of Cronus as he slept in his cave.

In corroboration of this we may note that Pliny the Elder, writing several years earlier than Plutarch (since he died in the Pompeii eruption of AD 79) would describe the islands to the west of Britain thus wise:

> There are those who record other islands: the Scandiae, Dumna, the Bergi, and Berrice, the largest of them all, from which the crossing to Thyle is made. One day's sail from Thyle is the frozen sea called by some the Cronian Sea [Sea of Cronus][16]

This might help us to identify that the 'largest island' is the Isle of Lewis and that Thule/Ogygia must be Iceland, with the coast of Greenland being the great mainland and the frozen sea beyond. We know that a thousand years later the Vikings would make similar transatlantic voyages, so why deny this capability to earlier northern mariners? Plutarch is here giving us more detail about the voyages to Thule that were already known to Pliny's older sources.

Those who might wish to consider all this as a merry fiction based on Plato and Pytheas should note that while Plutarch *could* have known the workings of the Celtic calendar; and *could* have and equated it with the calendrical details in Plato's narrative, there is no reason to believe that Plato or even his Egyptian source knew that they had a working calendar. It is also significant that we find the Neolithic stone circles at Callanish on the Isle of Lewis and Stenness on Orkney which could have been used, as part of their ritual function, to track the thirty-year correspondence of Saturn and the Moon. Plato or Solon could not have known the calendrical significance of these circles in order to construct a fiction about voyages to America; and as we have seen, their contemporaries didn't even know of the existence of the islands – or even that there was an ocean to the west.

And so to focus again...

We are examining here the origins of the ancient geography and ethnography given by an Egyptian priest to Solon around 600 BC, based on a story that was supposedly set down in writing at some time before 3000 BC in the Egyptian Delta. The arguments in support of this date we need not revisit.

A written account stabilises history at the time of writing, but an oral history or legend is constantly updated by the contemporary understanding of the people who propagate it. We may see that the knowledge of a large island somewhere in the Atlantic was preserved in North African legends both as a place of ancestry and as the abode of their gods. It was remembered as a green and lush island with a central plain and rivers. One version was very early written down and that account also remembered a catastrophe as related to Solon; other versions, presumably orally transmitted, continued to suppose that the island still existed out in the Atlantic Ocean opposite Spain.

The memory of this lost island with its fertile plain and rivers degraded to legend and then to myth; and we find it recorded in the Egyptian Book of the Dead as the rectangular Elysian Fields, their paradise of the dead. It was less precisely recorded in the Greek versions of these same myths; and also in the Phoenician, which became the basis of Carthaginian myth; and again in the Roman myths of the Underworld. All of these believed this sacred place to lie somewhere in the unknown west.

As Phoenician and Carthaginian traders expanded beyond the Straits of Gibraltar they took with them this mythology and vague geography of the Atlantic coast. There they met and traded tin and amber with mariners from Britain and Gaul who had first-hand knowledge of the real Atlantic islands; but they protected their trade monopoly by perpetuating the vague geography to confuse Mediterranean rivals. We see in the geography (before it was clarified by the exploration of Pytheas) the confusion of these myths with an island opposite Spain as one of the sources of the tin ore. This source was Cornwall at the tip of a peninsula lying 'opposite' the north coast of Spain and this would match the description in *Critias* of a peninsula of the lost Atlantic island that faced Gades.

It is only when we look to the few sources of early geography that we have from Greek and Roman writers who actually studied the history of the Atlantic coast that we find clarification. In the fragments that survive from Procopius, Posidonius and Pytheas; and the sources of the unknown Marcellus, we find the ancient beliefs of the inhabitants themselves, which believed the west of Britain to be the land of the dead to which the souls of the dead were ferried. They also record that people on an island at the centre of Britain and Ireland still remembered an ancient period when it had been the centre of a great empire. So where was this central island?

Notes and References

[1] Herodotus, IV,45

[2] Herodotus, III,113-4

[3] Hecataeus of Abdera wrote in the fourth century BC and wrote a book about the Hyperboreans; he should not be confused with Hecataeus of Miletus who was cited by Herodotus; only fragments survive from these authors.

[4] Diodorus Siculus, V, 21, 3-6 and V,22

[5] Oldfather, C. H. *Diodorus Siculus: The Library of History*, Loeb Classical Library, Harvard Univ. Press (1994). The reference is in a note on page 37 of his translation of Book II;

[6] Diodorus Siculus, II, 47, 4-6

[7] I have investigated these astronomical links more fully in *Under Ancient Skies*.

[8] Plutarch, The Parallel Lives, Sertorius and Eumenes, VIII

[9] Diodorus Siculus, V.20

[10] Ling, Johan; Stos-Gale, Zofia; Grandin, Lena; Hjärthner-Holdar, Eva; Persson, Per-Olof (2014), *"Moving metals II provenancing Scandinavian Bronze Age artefacts"*, Journal of Archaeological Science, **41**: 106–132, doi: 10.1016 /j.jas.2013.07.018

[11] Diodorus Siculus, V.22

[12] Strabo, Geography 4.5.1.

[13] Diodorus Siculus, V.38, 4-5

[14] https://romaninscriptionsofbritain.org/inscriptions/662

[15] I have elaborated on this in *Atlantis of the West* and *Under Ancient Skies*, where a reconstruction of the Gaulish Coligny calendar based on 11-year and 30-year cycles may be found in the appendices.

[16] Pliny, Natural History, IV, 104

5

An Island at the Centre

For many years, it was a common joke in Australia that the Aborigines were taught at school that they were discovered by Captain Cook! In fact they could tell vivid legends about the Dream Time and tales of extinct creatures that could not have existed for thousands of years; they even preserved legends about the advance of the sea around their coastline. Similarly, the indigenous people of Britain and Ireland did not need any invading Romans to tell them their own history and geography. They always knew that there was no such place as Thule, despite it being so well-known in Rome and Byzantium; and they preserved many legends of their own about the sea coast and of how the islands were first populated.

We have seen how awareness of what lay in the western ocean would become progressively more well-defined in the minds of Egyptian and Greco-Roman historians; and how in parallel with that, older recollections of an island in the Atlantic Ocean degraded within their mythologies to become barely distinguishable from the Underworld: the land of the dead. We must now focus-in to see what remains of ancient history and mythology from sources that are native to Britain, to Ireland and the people of the nearby Atlantic coast. We may then seek correspondences with the trail of evidence that brought us here.

Celts and Other Britons

The premise of a Celtic 'invasion' of Britain and the origin of the Celtic languages – the p-Celtic Britons (the Welsh) and the q-Celtic Irish has long been something of a 'sacred cow' in early-historical and archaeological dogma dating back to the nineteenth century and before. Similarly, it was only acceptable to whisper that the origins of their priestly caste, the Druids, in

Britain & Gaul might be earlier than the Iron Age, when they are first mentioned in Roman sources. To question these core beliefs would invite instant academic scorn and derision. As with Egyptology: so much rested upon the weight of eminent opinion citing earlier entrenched opinion. The intrusion of DNA evidence is at last beginning to unravel these old certainties and we may now look with more scepticism at what our older text books say.

The earliest literary references to Celts do not differentiate them from the Germans. We find in the reference of Herodotus that there were Celts around the source of the Danube, where by early Roman times we find only Germans. We know from various sources that other distinct nations had earlier existed to the west of these Celts: the Cynetes and Hyperboreans to name but two. By early Roman times all of these had been absorbed within a conquered empire or culture-province subject to a Celtic hegemony; and therefore the Mediterranean countries viewed them all as the same group of undifferentiated 'barbarians' to their north. Strabo considered all the Celts to be 'war-mad', quick to battle but otherwise a simple-folk!

It may be that within this culture-region the Celtic languages were able to expand more widely; but also the customs and religion of the people along the Atlantic coast could spread among the Celts; but here is not the place to re-evaluate long accepted doctrines about the Celts and their languages, unless it touches directly upon our subject. Suffice to accept that the languages that we call Celtic, and Celtic ethnicity, are not at all the same thing. The people of the regions and islands that we today consider Celtic because of their language are all situated in regions that were to the west of the ancient Celts; and those people who were once called Celts are now speaking German or Romance languages.

From the brief descriptions of Julius Caesar and Diodorus Siculus we learn that most of the tribes in the interior of Britain considered themselves to be aboriginal to the island and that for the most part they co-existed in peace. They were in no sense ethnic 'Celts' but must have converted to the Brythonic (pre-Welsh) language at some point between the date of the earliest references to Hyperboreans in an island beyond the Celts (i.e.

before about 800 BC) and the date of the earliest geography in the Greco-Roman sources. In Caesar's introduction we find:

> The interior of Britain is inhabited by tribes who claim on the strength of their own tradition to be indigenous to the island; the coastal regions by immigrants from Belgic territory who came only to pillage and make war...[1]

Despite this very clear differentiation between Continental Celts and the Britons by sources of history much closer to the events than us, this did not deter Celtic specialists from seeing nothing but Celts throughout Britain and Ireland.

For Ireland, the classical sources offer us even less to assist. Most Celtic scholars would view the arrival of Gaelic speakers in Ireland as earlier than the arrival of British-speakers. The old idea of successive waves of goidelic-Irish speakers arriving in Britain and Ireland, overridden by a later wave of British-speakers (the Belgae) is no longer current thinking among linguists; rather they had a different point of origin. Language conversion requires only the influence of a ruling-elite for a sufficient period of time, rather than an influx of population. Suffice to state that when we examine the earliest legendary pseudo-history in the Celtic myths, we are hearing ancient stories that have been translated from whatever language was spoken by the indigenous inhabitants in Bronze Age and Neolithic times

By 58 BC when Caesar used the intrusion of the Helvetii into the Roman province as an excuse to conquer all of Gaul, we may see that the Gauls composed a number of distinct nations under the general assemblage of Celts. The name of Celtica, or Gaul was applied to the two older Roman provinces, Cisalpine and Transalpine Gaul – the former Greek colony of Massilia. To the north were the tribes of the Lugdunensis occupying what is now central France and Switzerland; and beyond them the Belgae extended from the Rhine westward to the tip of Brittany. The Aquitani, occupying the modern region of Aquitaine, were probably Basques – but Caesar conquered them anyway! Beyond the Pyrenees we find again the Celtiberians who had established themselves in Galicia during the earlier Celtic expansion.[2] We need not pursue this ethnography beyond that necessary to establish the ground that we stand upon.

The Sacred Island of the Druids

In Roman descriptions of the Celts we find enigmatic references to their philosophers: the Druids. These were one of the ruling classes in the stratified society of the Celts. Their precise role equated more to that of wise judges, rather than priests, whom Strabo tells us were called Vates; along with Bards who were the guardians of the history and traditions, akin to the poets of the Greeks. The Druids would never permit their knowledge to be written down and initiates might spend as much as twenty years to learn their skills. The principal details of relevance for our purpose are summarised in Caesar's own commentary of his Gallic conquest:

> Of the two [ruling] classes mentioned one is the Druids, the other is the Knights. The former preside at religious ceremonies, administer...sacrifices, and advise on religious matters...

And of the origin of the Druid order:

> It is believed that the druidic system originated in Britain and was then brought into Gaul. There it is that those who wish to make a more detailed study of it usually go to learn.

On the matter of sacrifices he says:

> ...those affected by serious diseases or who fear the dangers of battle...offer human sacrifices, and they employ Druids to supervise this. They believe that unless one life is given for another, the will of the immortal gods cannot be influenced, and they also organise sacrifices of the same kind on behalf of the state. Others use huge figures, the limbs of which, woven out of branches, they then fill with living men. These are then set alight...

Most people will have heard of the 'wicker man' even if they know little else about Druids; the human sacrifices would typically be criminals or prisoners, but innocent persons would do just as well. Interestingly we are told:

> The Gauls affirm that they are all the descendants of Father Dis, and they maintain that this is the belief of the Druids.

Therefore we may presume that, if the order originated in Britain, then the Britons too claimed descent from Father Dis; this was a colloquial name for Roman Pluto or Greek Hades: the king of the Underworld. If the British tribes were indeed autochthonous then it would imply that this was a most ancient belief of the native tribes rather than of any later immigrants. Of the principal Celtic gods we are told:

> They worship Mercury [Hermes] first of all…After him they revere Apollo, Mars [Ares], Jupiter [Zeus], and Minerva [Athene], about whom they hold similar ideas as do other races.

We therefore see the equivalence of the gods and their powers between east and west, supportive of an ancient common origin underlying all these beliefs.

All of the Roman commentators of Caesar's era refer back to the lost history of Posidonius (135-50 BC). He of course had his own lost sources such as Ephorus, but we may be sure that he personally visited Gaul prior to Caesar's campaigns; and possibly also Britain. He remarks on how he was at first disgusted by the way that the Celts would hang the severed heads of enemies above their doors, but after seeing it so many times, became accustomed to it.[3]

Most of our detailed knowledge of Druids is therefore pertinent to Gaul and we cannot be sure that their roles and deities were identical throughout Britain and Ireland. Some of their practices were considered cruel and inhuman even by Roman standards. Augustus excluded Roman citizenship to those who performed Druid rites, which would be extended to a complete proscription by Tiberius. It was only after the conquest of Britain that Claudius would order that Druidism be completely abolished.

However, the wisdom of the Druids is clear from first century historian Pomponius Mela who tells us that even after the abolition of their 'savage customs', traces of their teachings remained and continued to be taught in secret forest glades. He says that they claimed to know the size and shape of the world, the motions of the stars and planets; and the wishes of the gods.

Secret knowledge therefore lay at the heart of the Druids' power over mere tribal kings, but it was also their weakness. Unless passed-on to their initiates by training, the ancient knowledge would be lost for ever – and after the Roman conquest much was indeed lost when so many were slaughtered and their centres of learning shattered. A few hundred years later, that which remained would be expunged by an even more effective slayer of ancient history – conversion to Christianity. This time the eradication would extend even to the unconquered north of Britain and to Ireland.

For a long time, archaeologists would frown upon any attempt to link Druids with the so-called pre-Celtic culture of the Neolithic megalith builders. This again derives from those old notions of a Celtic 'invasion' and the prejudice that every advance of culture could only have diffused to the west from the civilized east. The most obvious reason to question this is that the Druids' philosophy and astronomy, as we are plainly told, originated in Britain. The astronomical alignments at Neolithic stone circles and other aligned monuments display continuity from earlier times. The most compelling link is with the five-year calendar found at Coligny in Gaul and the 5 + 6 year cycles implied in Plato's narrative. All such calendars evolve conservatively from ancient roots.[4]

A number of references would strongly associate the Druids with the island of Anglesey or Ynys Mon as the modern Welsh call it. Caesar's introductory description of Britain continues:

> ...Another side [of Britain] faces Spain and the west. In this direction lies Hibernia [Ireland] thought to be half the size of Britain and as far from it as Britain is from Gaul. Midway between the two is the island called Mona and in addition it is thought a number of smaller islands are close by, in which...there are thirty days of continuous darkness around midwinter.

We see repeated here the persistent geographical error that the west of Britain faced towards Spain. Here, Mona implies the Isle of Man, but Pliny and Tacitus used *Mona* for Anglesey and we find *Monapia* for the Isle of Man.

We encounter the island of Mona again in accounts of the Roman conquest. It was to the tribes in Wales that Caratacus retreated to rally support after the defeat of the southern tribes in the first phase of conquest. In AD 60 after first conquering the Silures in south Wales, Suetonius was in the process of subduing the Ordovices in Anglesey when he had to hasten south to confront Boudicca's revolt. It was to be only a brief respite, for in AD 78 Agricola would return to complete the conquest. Remnants of older British traditions would be remembered during the years of the Roman province; and persist within the unconquered parts of northern Britain, but Rome would never entirely subjugate the north.

Writing of the campaign in Anglesey, Tacitus again remarks on the abhorrent practices of the Druids, who would read omens in the entrails and contortions of a sacrificed victim – ritual murder as the Romans viewed it. We must place this alongside any merit that we assign to Druids for their wisdom in astronomy and philosophy. But were their practices so much worse than the Roman crucifixions? Were their sacrifices any worse than those of the Mayans; or the hanging-drawing-quartering of traitors in more recent centuries? Precisely what aspect of Druidism so appalled the Romans remains elusive. It may be that their beliefs were not so different from Roman religion, rather that their similarity in many respects was considered a parody and desecration.

The prominence of so many women in Druid rituals and in positions of power was another element totally alien to Roman society. Tacitus records that Suetonius encountered women in black robes, with long hair and flaming torches, who terrified the Roman soldiers with their wailing and curses until urged forward by their commander. There was great slaughter as the Druids were consumed by their own flames. Suetonius completely destroyed the sacred groves or 'nemetons', which were always situated in forest clearings. These we may equate with an archaic shamanist strand of belief, whereby spirits and demons were believed to reside in sacred sites, such as wells, springs, rock formations and even in long-lives trees. One may perhaps therefore consider druidism as a highly-evolved form of the shamanism found throughout northern Europe and Asia.

The modern archaeology of Anglesey will reveal many interesting features in the landscape. The island holds numerous Neolithic monuments, such as Bryn Celli Dhu, described as a chambered tomb, aligned to the midsummer solstice, and dating from around 3000 BC. Other 'tombs' such as Tŷ Newydd Burial Chamber, classed as a dolmen, may date from the earlier Neolithic; and the Castell Bryn Gwyn henge of Bronze Age date. Any of these sites could have been sacred sites to later Druids, though it is difficult to be precise as to where the later sacred groves actually were. In 2017 a series of early-Neolithic houses dating from 3800-3600 BC were excavated at Llanfaethlu along with human remains; and pottery classed as 'Irish Sea ceramic ware'.[5] The continuous occupation of Anglesey back into the Neolithic era is thus well established.

With all respect due to the modern inhabitants of Anglesey, it is difficult to see quite why their island should merit so much attention from the Roman authorities, and so much prominence in our written sources, unless it was indeed the sacred island of the Druids: the very core of their cult. Why was this location so holy to them? We may compare it to Mecca or Jerusalem as a place of pilgrimage for the faithful. So long as its influence persisted then resistance to Rome would continue. We may compare the Roman campaign with the better documented revolt of the Jews and the relentless drive of Vespasian to suppress their religion in Judaea. Clearly, Vespasian, who knew Britain first-hand as well as he knew Judaea, was prepared to employ similar methods. We may see therefore why he sent out the expert Demetrius to report on their religion. We may even see the reasons why Claudius determined that Britain must be conquered, to eliminate this source of agitation from Gaul.

There are a few other clues about Mona in the chain of evidence that must be included before we can move on to the mythology. We know from Pliny that the Druids used a lunar calendar, which we recognize as the Gaulish Calendar of

Coligny; and that they also measured ages of thirty years:

> Mistletoe…is gathered…above all on the sixth day of the
> moon (it is the moon that marks out for them the beginning of
> months and years and cycles of thirty years)…[6]

On this special day a bull-sacrifice was performed – but it is not
clear whether this was monthly or only each thirty years. The
significance of this thirty-year period as marking the orbit of
Saturn was given to us by Plutarch in his description of the
voyages to Ogygia. From these few astronomical details it is
possible to reconstruct how the calendar must have worked.
The astronomical knowledge of the Druids is further confirmed
by another remark of Pliny concerning the Arctic days and
nights:

> Pytheas of Marseilles writes that this occurs in the island of
> Thule, six days voyage north from Britain and some declare it
> also to occur in the island of Mona, which is about 200 miles
> from the British town of Colchester.

We need not debate the precise significance of this apparent
geographical error of latitude, but it can only be a survival from
the astronomical knowledge of the Anglesey Druids
themselves; for who else would know or care?

Therefore, we are entitled to ask: was Anglesey with its
colleges of Druids and Bards that 'central island' which
remembered that it had in ancient times been at the centre of an
empire extending as far as Africa and the borders of Egypt; and
what is the link between this central island and the submerged
island that Plato described?

Mysteries and Sacrifices

The geographer Strabo, writing some time before the Roman
conquest, remarks on certain sacrifices practiced by Celts on an
island close to Britain:

> He [Posidonius] says that there is an island near Britain on
> which sacrifices are performed like those sacrifices in
> Samothrace that have to do with Demeter and Core…Ephorus

in his account, makes Celtica so excessive in its size that he assigns to the regions of Celtica most of the regions, as far as Gades, of what we now call Iberia; further he declares that the people are fond of the Greeks, and specifies many things about them that do not fit the facts of to-day.

However, we are not told with certainty that this island was Mona. The reference is to a single island. It could refer to Ireland, but surely he would have named it as such; or it could have been one of the Scottish islands. One might despair that our sources are so often ambiguous and that so much detective work is needed to establish the facts. Ephorus wrote in the time of Alexander the Great before 330 BC and so he is describing the situation earlier than that; he was greatly influential to both Diodorus Siculus and Polybius.

By one of those many co-incidences, you may recall that Samos, or Samothrace, was one of the Aegean islands which Myrina and her Amazons visited and (according to Diodorus) it was she who instigated the first sacrifices on that island. It may therefore be worthwhile to look in more detail at that quotation. The island became known as Samothrace due to some Thracians from the mainland who later settled there:

> ...the Mother of the Gods, well pleased with the island, settled in it certain other people, and also her own sons, who were known by the name of Corybantes—who their father was is handed down in their rites as a matter not to be divulged; and she established the mysteries that are now celebrated on the island...[7]

We may posit, if the ceremonies were indeed similar to those in Britain, that their undisclosed ancestral 'father' was Hades, just as the Gauls and Britons would trace their ancestry to the king of the Underworld.

Despite over two thousand years of debate, no-one is quite sure what the mysteries of Samothrace actually were; they really are a mystery. It is difficult enough to extract ancient history from our meagre sources; the more so when the ancient people themselves set out to hide it. Herodotus, if you wish to believe him, offers further information: that the Samothracians learned their rites from the earlier inhabitants, the Pelasgians, before they taught the mysteries to the Athenians.[8] If the fertility rites of Demeter were indeed similar to those at Eleusis near Athens then we may view it as a celebration of the annual fertility of crops in springtime. Core was Persephone, the daughter of Zeus and Demeter. At the rites in Athens her name was unspoken and hence she would simply be referred to as Core, which implied: the maiden or virgin. We may therefore view Demeter as another variant of the earth-mother fertility goddess found throughout the ancient Mediterranean world. The Egyptians would equate Demeter with their own goddess: Isis and of course their equivalent of Hades was Osiris – he whose name Herodotus declined to speak.

The Eleusinian mysteries recalled the kidnap of Persephone from her mother Demeter by Hades, the king of the Underworld, who carried her off prematurely to his realm. Demeter is then said to have gone in search of her daughter and was eventually reunited with her.

The symbolic 'rebirth' of Persephone became an annual spring festival during the Hellenic era and in later times also at Rome. So we may envisage an analogous annual ritual taking place in Britain.[9] The question would then be whether that correspondence is due to ancient colonisation of the coastal Mediterranean by people from the Atlantic coast, as is suggested by the myth of the Libyan Amazons; or whether we should view it rather as a rite brought west by traders and pilgrims from the Aegean during the later Bronze Age. Recall again, Plutarch's story of Greek colonists living on the Scottish islands.

Rather, it is the myth of the Underworld preserved by Procopius that gives us a clue to why Mona-Anglesey was so important to the Druids. Their cult preached that the soul was immortal and would be reborn in another body in another world. Classical authors would liken it to the philosophy of Pythagoras. Perhaps the most succinct statement of this comes from the poet Lucan (mid-first century AD):

> They alone are granted the true knowledge...of the gods and
> celestial powers; they live in the furthest groves of the deep

forests; they teach that the soul does not descend to Erebus'
silent land, to Dis' sunless kingdom, but the same spirit
breathes in another body.[10]

Again, we see why sometimes only a quotation can show that
the evidence is not being 'forced' – it really is there in the
ancient sources. The island of Mona was sacred to the Druids
because it was that place to the west of Britain that was
believed to be closest to the Underworld – the Elysian Plain.
This was the 'other world' into which the immortal souls were
thought to be reborn.

The Irish and the Fomorians

When we look at mythology from Ireland we are seeing a quite
different kind of degraded history to that found in the Greek
and Egyptian sources. Authenticity is conferred by how close to
the events a story is first crystallised in written form. This
preserves events and characters closest to proper sequence
although we may not always be able to assign them a precise
date. Even for Greece, the verse in which half-remembered
myths were recorded dates from around 800 BC. However for
Britain and Ireland there was no writing; indeed to write-down
the knowledge was forbidden. We are reliant therefore upon the
accurate oral transmission by generations of Bards. The
conversion to Christianity curtailed the tradition of the pagan
bards in the Celtic countries and the stories of origin and of the
Celtic Otherworld degraded into timeless folklore, to be
passed-on by story-tellers who no longer understood the
original significance of their content.

The written sources of tradition upon which we rely are
found only in late medieval documents and none predates
Christian conversion. As in the Greek poetry, we do not see in
the Irish stories a desire to preserve precise history; rather they
are constructed to entertain an audience with interesting fiction
about heroes with whom they were already familiar. As a
potential source of ancient history we might compare them
rather to that American western film with John Wayne at The
Alamo; real heroes are set in authentic places or made to relive
historical events, but in a purely fictional plot.

Of primary interest are the stories in the Book of Invasions within the Mythological Cycle, which may preserve something of the pre-Celtic beliefs. We are offered six invasion stories; although the source manuscripts are relatively recent, the gist was known to Nennius in the ninth century who would rationalise the point of origin for all the Irish colonists as Spain. The later versions of these same stories would bring the colonists as successive waves from Greece or the Aegean.

The first legendary invader was Partholon and his tribe. On Christian conversion, the myth of Noah was prefixed; as a relative of Noah he was refused entry into the Ark and upon consulting an idol they fled to Ireland to escape the flood. Upon arriving in Ireland the geography was not as we know it today; it had only three lakes, nine rivers and one plain, called the old plain. The island was still in a state of evolution as another lake burst into existence. They had to battle and drive off a race called the Fomorians (*Fomhóraigh*) who were already in the neighbourhood of Ireland. But all of Partholon's tribe died of plague and the island then lay deserted for thirty years.

The etymology of the name *Fomhóraigh* is interesting. It has been suggested from medieval sources that 'mur' the old Irish word for the sea is contained here, and together with 'fo' it implies something like 'under-sea-people'. Others would prefer to take it from 'mór' meaning big or giant – hence: 'under(world) giant'? The double meaning could have been taken both ways by an audience which may be why we so often see them described as giants. Either way, it is another interesting coincidence to add to the list. Incidentally *Fomhuiri* is the modern Irish word for a submarine.

The next colonists to arrive were the Nemedians, the Tribe of Nemed, of whom many were lost at sea on the way, or died of unknown illness.[11] The story says they were Greeks, originating from Scythia, who came via rowing across the northern ocean. They too found Ireland in a state of flux, with more lakes bursting out. They cleared twelve plains and built towers, before they had to battle with the Fomorian kings Morc and Conann who ruled from a tower on Tory Island off the coast of Donegal. They demanded tribute of two-thirds of the colonist's corn, their milk and their children. Not surprisingly, the Nemedians rebelled and they sent messengers back to Greece to bring more warriors. Suffice to note that after a partial victory the tribes dispersed to live in other countries over the sea or went back to Greece to escape the plague; and the island again lay empty, this time for two hundred years.

The next people to arrive were the Fir Bolg, whom Nennius tells us occupied the Isle of Man and other islands nearby. These were descendents of the Nemedians, returning again to Ireland after two hundred years (and we must presume it was they who remembered the earlier extinct colonies) and they came in three groups and settled in different regions. At some later time, we know not how long, came the Danaans, or the *Tuatha de Danaan*, from whom: 'all the learned men of Ireland are descended'. These were yet another branch of the Nemedians, of whom we are told:

> After leaving Ireland...they settled in the northern islands of Greece...They learned druidry and many various arts in the islands where they were.

But unfortunately we go off our subject to follow the various comings and goings of Nemedians and Greeks.

The Danaans made an alliance with Balor the king of the Fomorians, cemented by a marriage to a Fomorian princess, from which the hero Lug Lamfada ('of the long arm') was born. The Fomorians would allow them to settle in Ireland. Danaans and Fir Bolg fought a battle, known as the first battle, at Mag Tured (English: 'Moytura') a plain in the north of Co Sligo, and the Danaans were eventually victorious; but still fearful of the Fomorians they would eventually agree to divide Ireland between them. The Fir Bolg were allowed Connacht and left the rest to the Danaans. It is From among the Danaans that we find later Irish heroes such as Manannan and the Dagda; and it is these semi-divine heroes who are chosen as leading characters to visit the Celtic Otherworld in the various later tales.

In the course of first battle, Nuada, king of the Danaans lost his sword-hand and hence disqualified to be king. So Eochaid Bres son of Elotha was chosen as king in order to preserve the alliance with the Fomorians. Now in this regard there is a tale that Elotha was a Fomorian prince; and that the birth of Bres came about from an Otherworldly visit by Prince Elotha to his mother. This is typical of later myths where the Tuatha De Danaan would become semi-divine in nature and flit between the real world and the Otherworld, but in these early invasion myths they are otherwise described like a race of purely human invaders.

Nevertheless, the Fomorians demanded tribute and would oppress the Danaans. To summarise here a long story, the Fomorians gathered a great army to enforce their rule upon the Danaans and collect their taxes. Thereafter any man who refused to pay his taxes would have his nose stricken-off. There then arose the heroic leader of the Danaans, Lug Lamfada, who was by now a young man. He is said to have returned from the 'Land of Promise' and brought with him Manannan's sword, his boat and his horse; and assisted by these magical weapons they would defeat the Fomorians. But Lug allowed the few survivors to return to their homeland and when Balor the king of the Fomorians heard of the defeat he gathered a great army and fleet and set upon revenge. Again, to summarise a complex story, the great battle, known as the second battle, took place again at Mag Tured; here Balor was killed and the Fomorians routed, this time finally and we hear no more of them.

After another unspecified period of years there arrived the Milesians, who are typically identified with the modern Irish. It is far from clear whether this was a real invasion or a later rationalisation to explain a change of culture. The story treats the Milesians as real people, rather than semi-divine beings like the Danaans. They defeated the Danaans in battle; but the legendary account says that the Danaans were permitted to live underground, or in the *Sidhe* mounds; this name (English: 'Shee') originally referred only to the tumuli and associated ruling palaces, but over time it has also come to refer to the people themselves. It may be that they persisted, as subjugated races often do, absorbed among their conquerors.

All of these myths of invasion are quite timeless and we may only guess at era, or indeed which of the invaders brought with them the Irish language. The supposed comings and goings from Greece hint at an early date, long before the Greek Dark Age; certainly before the Phoenicians blocked the sea routes and Greece forgot about what lay in the Atlantic Ocean. We may perhaps liken these invasion myths to the early colonisation of America: a few ships of early settlers are remembered, followed by others unrecorded, with the Fomorians taking the role of the 'Indians'.

There is no myth that any part of the Irish population believed themselves to be aboriginal to the island, but the Fomorians are the closest that we come. In the account of the second battle at Moytura they are thus described: 'never came to Ireland an army more horrible...men from Scythia and men out of the western isles'. We are told that after the first battle of Moytura the surviving Fir Bolg sought refuge among the Fomorians, 'and settled in Arran, Islay, and in Man and Rathlin'. So this again tells us from where they originated. In the myth about the birth of king Eochaid Bres we have a careful description of his Fomorian father: Elotha:

> ...she saw...a man of fairest form. Golden yellow hair was on him as far as his two shoulders. A mantle with bands of golden thread was around him. His shirt had trimmings of gold thread. On his breast was a brooch of gold...Five circlets of gold adorned his neck, and he was girded with a golden-hilted sword...

And he arrives and returns across the sea in a shining silver vessel; tall and otherworldly he may have been, but this was no ugly barbarian giant.

When the later invaders arrived in Ireland they must have discovered all around them the dolmens and abandoned fields of an earlier culture, as well as the later monuments, and would have asked the inhabitants who had built them. These 'faery mounds' the *Sidhe*, would therefore come to be associated by later Irish with the people who preceded them, or we may prefer that the earliest monuments were actually built by the ancestors of the Fomorians and it was they who preserved traditions.[12] It is primarily due to this faery-association that we see the degradation of the Fomorians to become mere giants in 'fairy stories' for children; becoming somehow even less believable than a myth.[13] In some academic studies of Celtic myths one will even find the Celtic Otherworld referred-to as 'fairy-land' or in such-like dismissive terms. This does not assist the study of the underlying historical origins and has surely prohibited serious consideration of valuable evidence.

Archaeology assigns the building of the so-called faery-mounds, the burial mounds at *Bruig na Boinde*: the bend in the Boyne, to a date around 3100 BC. These comprise the Newgrange mound and the lesser mounds at nearby Knowth and Dowth; with the earliest aligned stone circles being somewhat later constructs; so if one would wish to associate the Danaans with the building of these mounds then the correct era to place them would be: third millennium BC – the Late Neolithic. The early Irish invasions would therefore fall somewhat before this time, with the Milesians at some much later date. But as always with legends one cannot be sure that the details are in the correct sequence; with myths and legends muddling is the norm.

As for the presence of potential mythological 'fossils' to authenticate the invasion myths, useful pointers are the references to the colonists being wiped-out by plague and unknown diseases. This rings true by comparison with our knowledge of later European colonisations of the Americas, where it was instead the natives who were infected rather than the colonists. An ancient storyteller could not have known to invent such a detail.

There are so many points of interest in the Irish myths but one cannot follow them all. One detail in *The Fate of the Children of Tuirenn* stands out for the number of useful clues that it offers to the ancient past. In the story of the Tuatha De Danaan and their struggles with the Fomorians, we are told:

> Then the chief men of the Fomorians went into a council, namely Eab the grandson of Net and…[six others]…and Lobais the Druid and Liathlabar the son of Lobais; and the nine deeply learned poets and prophetic philosophers of the Fomorians, and Balor of the stout blows himself…

This is another of those details that fulfils the definition of a mythological 'fossil'. Firstly, it is not strictly needed in order to tell the story of the battle; we don't really need to know that the Fomorians possessed Druids and the story-teller is explicit that there were nine poets and philosophers. Precise numbers in a myth are always a clue that it is a fragment of real history. This matches closely the behaviour of later Gaulish Druids given by Diodorus, who says that they would sometimes intervene in disputes between kings. Note also the presence of bards to record the events.

Most useful however, is the simple detail that the Fomorians possessed wise Druids (as the Irish understood them). In the earlier account of the invasions of the Nemedians, we are told that before the battle between the Nemedians and Fomorians at Conan's tower a battle was begun 'between their Druids and another between their Druidesses'. This does not fit with metaphors elsewhere that would turn the Fomorians into ugly giants and pirates. These were people of wisdom with a rich culture of their own. It gives us the clue that their homeland was northern Britain, around the Irish Sea and the western islands. While we cannot assign a precise date, the Fomorians are found only in these early stories of invasion. It gives us further confirmation that the beliefs and traditions of the Druids evolved very early among the aboriginal population of Britain and Ireland. We are here perhaps given a glimpse into the society of the pre-Celtic population who built the burial mounds and the stone circles. Were these also the seafarers who would attempt the long voyages to Ogygia?

The Celtic Otherworld

In many of the Irish myths we see a belief in the existence of an 'otherworld' and often, a visit to this Otherworld is the entire purpose of the story. It is encountered under several names, which may perhaps refer to different regions. The most well known is surely: *Tír na nÓg* ('the land of youth') found in the story of Oisin and Niamh, where the hero is led into the Land of Youth by a beautiful princess. In another story we find *Mag Mell* ('the plain of delight'), where the hero Connla is similarly invited to the 'Land of the Living' by a woman who casts a love spell on him. The most revealing of its names is *Tir fo Thuinn*, which means 'land under the water' – giving us the clearest indication that it is a place beneath the sea. Other names clearly perceive it as being a great plain: a delightful colourful land of beauty and perfection with towers and palaces; and there is no cold winter in the Otherworld, neither seasons nor storms.

In the various adventures of the heroes in the Otherworld we never perceive a dark or gloomy place like the classical Underworld. Rather, we see a kind of parallel idealised world into which the characters may venture or be invited by one of its inhabitants, often a beautiful woman. Usually, like Odysseus, the hero returns (for how else could they tell the tale?). In many ways it more closely resembles the Egyptian version of the Elysian Fields: that happy place, where the righteous dead could spend an eternity free from toil. It echoes again the Druid's belief; that the deserving souls were not truly dead, but occupied another body in another world, like our own yet somehow kinder.

In the story of *Ruadh in the Land under the Wave*, probably composed before the eighth century, Ruadh the king of Munster is on his way with his fleet to meet the king of the Norwegians when his ships become stuck in mid-ocean. They throw treasure into the sea to placate the sea-god, but he will not release them. So Ruadh dives into the water to see what is holding them and immediately finds himself in a great plain beneath the sea. There he is greeted by nine lovely women who admit that they have trapped his ships in order to draw him down. We encounter this theme again in one of the best known stories: *The Wooing of Etain* where we have another version of the plain, for we are told that the plain of Ireland was but a desert compared to the beauty of the Great Plain.

In: *The Voyage of Bran* we find a more detailed description of the undersea plain and where it was thought to be. The hero is King Bran son of Febal, who is lulled to sleep by enchanting music while out walking. On returning to his castle, a woman described as wearing strange clothing (indicative that she is an Otherworldly spirit) appears in the castle and sings a poem of fifty quatrains to the assembled gathering of kings and princes.

The poem tells of a 'distant isle' around which 'sea-horses glisten' and before vanishing she invites Bran to visit her land.

So Bran sails off across the sea and after two days at sea they encounter a strange man riding across the waves in his chariot. The description alone is enough to establish him as Manannan Mac Lir, or 'Manannan son of the sea' who is the Irish sea-god. He sings to Bran another poem of thirty verses taunting Bran that he is unable to see his realm in the waves below.

> Bran deems it a marvellous beauty
> in his coracle across the clear sea:
> While to me in my chariot from afar
> it is a flowery plain on which he rows about.
>
> That which is a clear sea
> For the prowed skiff in which Bran is,
> That is a happy plain with a profusion of flowers
> to me from the chariot of two wheels.
>
> Bran sees
> the number of waves beating across the clear sea:
> I myself see in Mag Mon
> Rosy-coloured flowers without fault.
>
> Sea-horses glisten in summer
> As far as Bran has stretched his glance
> Rivers pour forth a stream of honey
> In the land of Manannan son of Lir

Manannan taunts Bran in succeeding verses that the waves he sees are actually the land; that the leaping salmon are really calves and lambs; that the plain is *Mag Mell* on which horses dash around; his coracle has floated over an orchard of fruit trees; and so it goes on.

The name *Mag Mon* (or 'Moy Mon' for the English speaker) is usually taken to mean 'the plain of sports', another name of the Otherworld. It is significant that the story is recorded in a poem, which may confirm its ancient origins; this being the form that was more easily learned by the bards who would recite it orally to an audience.

Manannan Mac Lir is recalled in Manx legends as the king after whom the Isle of Man is named (or perhaps it was the other way around). We may view him as a demi-god, rather like the deified Caesar; a great ruler so-honoured and deified after death. Manannan's era is uncertain. According to *Cormac's Glossary*, in the tenth century Book of Lecain, the real Manannan was remembered as a merchant and Druid who traded from the Isle of Man; and who had an uncanny ability to predict the weather and seasons.

The internal evidence of the myth itself is plainly telling us that the undersea plain was believed to hide in the vicinity of the Isle of Man, for this is why Manannan is employed in the story, specifically to hint at the location. The pagan bard who composed the poem could only suggest the location in a cryptic manner, for it was part of the 'secret' knowledge. The Christian bard who eventually wrote it down no longer understood the obscure references, for by Christian times the pagan Otherworld had been entirely displaced by the Christian heaven. It is unfortunate that modern Celtic scholars have not been brave enough to accept what the myth actually says instead of going off into fancies of their own. For example Tom Peate Cross and Clark Harris Slover (whose translation is followed here) would choose the safe option and remark on its similarity to Homer's Odyssey, yet advocate that it was 'probably based in large part on fantastic stories brought back by sailors who had ventured far out into the Atlantic long before the discovery of America'.[14]

As we may see above, the Otherworld of Irish myth was never a vague place like the classical Underworld, allowed to exist somewhere at the edge of the world in the unknown western ocean. From the various references we may even see clues that it was viewed as co-existing in parallel with the living world, or more plainly still continuing beneath the Irish Sea. Later belief would connect it with the various Neolithic monuments, the Faery mounds, as the symbolic entrances to the Otherworld; perhaps telling us something about their original ritual purpose.

Most Irish scholars would concur that the detailed stories of Tara and the high kings of Ireland (except where the heroes stray into the Otherworld) preserve details of the Iron Age society of the island, perhaps approaching times parallel to the Roman incursions of Gaul and Britain. These stories of early Irish history and identity are not our primary focus. However, it is fair to note that in the Irish pseudo-history we do not observe the cruel Druid sacrifices that Roman commentators describe in Gaul and Britain. Irish Druids seem rather to have behaved like magicians or shamans. It may be that their worst deeds were censored by the later Christian bards.

When Lakes Burst and the Sea Widened

We also find in the Irish myths numerous references to the geography of Ireland and the changes that have affected it. In the story of Partholon, we are told that in this remote time Ireland was desolate with only a single plain (i.e. a cleared agricultural region) referred-to as the Old Plain. It may recollect some ancient geography going back to the earliest arrival of Mesolithic people after the ice withdrawal, while the land was still rebounding from the weight of ice.

There are more of these curious references to the topography of Ireland. In the legend of Nemed, we are told that four lake bursts occurred within the space of nine years; and they are named for us: Loch Calin Ul Nialan, Loch Munremar of Sliab Guaire, Loch Dairbrech, and Loch Ainninn. Again, where we see precise numbers and names in a legend we should respect it as a genuine memory. These references are conferred with a ring of truth because the Irish bards could not have known anything about the rebound of land after removal of the ice burden, or even that there ever was an ice age; and perhaps there were other tectonic causes for changes to the water-table. We may also seek in the Celtic myths any memory of a flood catastrophe that could have affected the island and which forever changed its bounds.

Some of the surviving mythological tales are prefixed by the Biblical creation story and the Flood of Noah, with the usual formula tracing descent from the patriarchs. Whatever pre-Christian legends of the Flood the Celts and Irish may have held have been obscured. Pre-Christian flood legends may persist, however, in local place name stories.

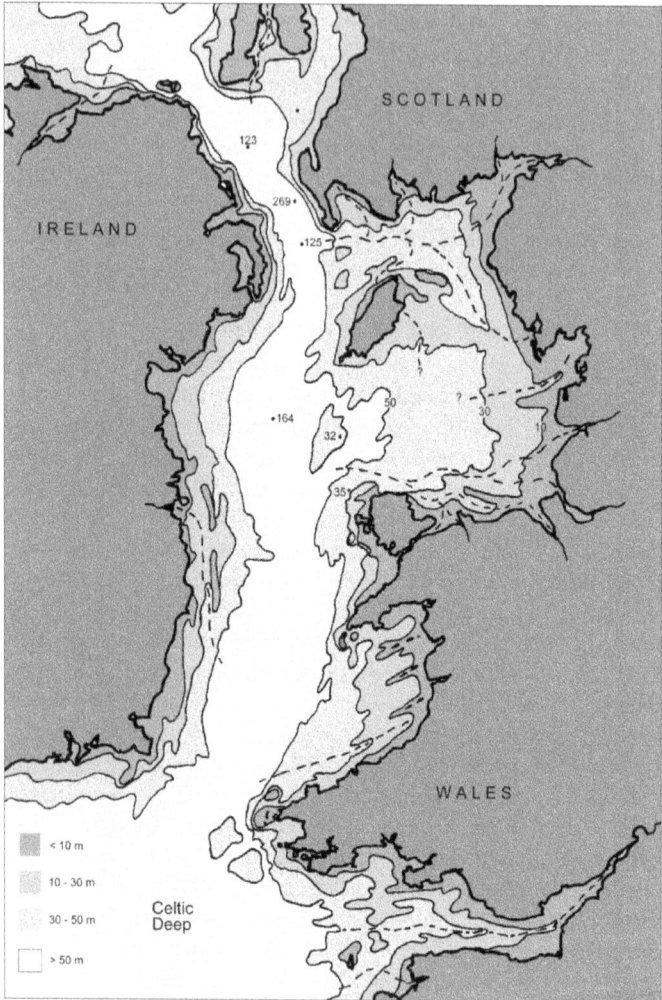

Figure 5.1 The 10, 30 and 50m submerged contours and submerged river channels beneath the eastern Irish Sea. This chart was originally included in *The Atlantis Researches* (1995). Based on UK Admiralty charts 2, 1121, 1826 and 1411.

From Glandore, Co Cork, there is the legend of *Tonn Clidna* or 'Cleena's Wave', a beautiful princess from the Land of Promise, who drowned in the first of three great floods, those of Clidna, Baile and Ladru. We are told: 'This was the time at which the illimitable seaburst arose and spread throughout the regions of the present world'. This and similar stories from Wales give us the evidence that the Celts viewed their equivalent of the Biblical Flood as the bursting or overflowing of the ocean, modelled on the bursting lakes with which they were familiar.

At various places around the coast of Britain and Ireland we find traces of submerged forests, which are often exposed only at low tide. These show stumps and sometimes fallen trees still in their position of growth, which were overcome by the sea and buried by sand before they could decay in the normal way. The term may also be used of any submerged peat layers overlain by a beach. Submerged forests are a world-wide curiosity because they cannot be created by a gradual inundation of the Sea. Young trees cannot germinate in a salt water environment but once inundated a mature tree might survive for many years; this implies that the woodland grew above the tidal range and was then rapidly overcome by the sea. Submerged forests could not be explained by a tsunami; for their creation requires a rapid but also a permanent submergence.

The most well known example is at Borth near Aberystwyth where the submerged forest has been dated between 3300-3100 BC and a similar forest loosely dated 4,000 years old was revealed at Ballinskelligs in Kerry, Ireland by the storms of 2014. In the Orkney Islands tree stumps near Stromness have been dated to around 6,000 years ago.[15] Submerged trees have also been exposed from time to time along the Cornish coast around Mounts Bay, dated loosely between 4,000 and 6,000 years ago. Tree stumps of similar age also underlie the Lancashire coasts between Blackpool and the Wirral.

The distribution of these forests from the far north of Scotland to Cornwall at roughly the same era tells us that the sea-level rise was not some localised earthquake phenomenon, but was likely part of the same wider causal event. Since trees can live for hundreds of years then a radiocarbon dated sample could give a date from any time during the lifetime of the forest. Radiocarbon dates therefore do not provide a precise date of submergence.

This rapid inundation was witnessed by the people who survived it and it gives support to the many legends; for related to this phenomenon is the folklore of sunken cities around the British coast. The best known legend is that of Cantrae'r Gwaelod, a lost land that is said to have been overwhelmed in Cardigan Bay, where we also find the dateable submerged forests. To confuse this picture, there are also legends of lost castles or cities around Penmaenmawr and St Davids. A Welsh *triad* in Latin translation, from Exeter, also describes three Welsh kingdoms that were lost to the sea: the kingdoms of Teithi Hen, Helig Ap Glannog and Rhedfoe. Teithi the Old is also mentioned in the Mabinogi of *Culloch and Olwen*, as he 'whose kingdom the sea overran', but we are not told where it was! This offers a further clue that these legends of submergence are ancient.

In the Brittany peninsula of France we find the similar legend of Ker-Is. This was the same region where Procopius records colonisation of Armorica (Brittany) by refugees from Britain in the sixth century AD. Therefore we cannot be sure whether the tradition of Ker-Is arrived with these colonists or has survived there independently from pre-Roman times.

One cannot omit here the legend of *Lyonesse* preserved among the Arthurian legends. The version recorded by William of Worcester in the fifteenth century says that to the west of Land's End there was once a prosperous land of many towns and fields, with a hundred and forty churches. The popular legend recalls that Lyonesse was submerged so rapidly that only a single survivor named Trevilian was able to escape by riding inland on a swift horse. There are also legends associated with submerged sea walls between the Isles of Scilly, which suggest that the islands were once linked as a single island.[16]

We must bear in mind that the pre-Welsh language, *Brythonic* was formerly widespread in Britain and was related to continental Gaulish. Therefore where we find such localised legends, then similar stories may also have existed in the north.

Procopius hints that the people of Cumbria and Strathclyde may have had similar legends of submergence, before their language was displaced by English. Those who have researched these lost-city legends in detail suggest that they evolved from a common source that remembers a single ancient inundation somewhere around the British coast. This memory was then localised to other districts in later folklore to explain some visible offshore anomaly. The possibility that an ancient inundation might have taken place all around the coast is simply too much for most researchers.

A Bridge Too Far?

Modern visitors to the Giant's Causeway stones on the coast of Northern Ireland will be told of the folklore; that they once marked the beginning of a causeway built by the ancient giants to link Ireland with Scotland. The other end of this mythical causeway was said to be the related volcanic outcrop around Fingal's Cave on the Scottish island of Staffa. Of course, we may be positive that the hexagonal columns are a natural prehistoric lava flow; but what is the basis of the folklore? The Irish name: *Clochán na bh Fomhóraigh* means something like 'the stepping stones of the Fomorians'; implying that the myth of a land-bridge was associated with these earliest inhabitants of the islands.[17] It gives us another clue that the real home of the Fomorians was in Britain; in the highlands and islands to the north.

The common folk-tale associates the causeway with the giant: Fionn Mac Cumhaill (or Finn Mac Cool to the English-speaker) who was challenged to battle with the Scottish giant Benandonner and so built the causeway to reach him. It's a delightful tale for children, but there are more solid references in mythology that hint at the existence of an ancient land-bridge in the time of the Fomorians.

For example, in the Welsh *Mabinogion* we find the story of Branwen, Daughter of Llyr (or daughter of the sea-god Ler). His Irish origins are betrayed by the retention of the Irish genitive case: 'of-Ler'). We hear that the hero Bran is able to sail across to Ireland from the Menai Straits – or if he preferred then he could instead walk to Ireland. At that era of the

Otherworld, Britain and Ireland were separated by only two rivers, for we are told:

> [Bran]...and the host of which we spoke sailed towards Ireland, and in those days the deep water was not wide. He went by wading. There were but two rivers, the Lli and the Archan were they called, but thereafter the deep water grew wider when the deep overflowed the kingdoms.[18]

Many translators have had difficulty in understanding how to translate this passage, but we may deduce that the events of the tale are taking place at a time before the loss of the land-bridge to Ireland.[19] The clause is surely included in the story solely to establish the era. Note the casual form of words employed and compare it to the similar wording found in the Atlantis myth. There is no earthquake, no convulsions, no deluge of rain, no volcanism; the sea *widened* and *overflowed*. Also, it spilled over not just the land, but over *the kingdoms*. Therefore it follows that it drowned cities and people. The storyteller feels no need to explain this extra-ordinary event to us; it is simply assumed that the audience will know all about it, just as if he were telling us today that the events occurred before Noah's flood.

It is another of those remarkable coincidences that the western part of the North Channel is the only place where the sea-bed is sufficiently shallow for a land-bridge between Ireland and Britain ever to have existed. On navigation charts, a ridge line may be traced between Malin Head in Donegal and Islay where the depth is only 20-30 metres, compared with depths in excess of 50 metres further east. It is actually the eroded terminal moraine of an ancient ice sheet and probably once stood much higher. The more surprising then, that the myth of a land bridge should exist in this place and not elsewhere; there are no such myths where the crossing is at its shortest (and deepest) between Larne and Galloway; or further south between Wicklow and South Wales. How did the ancient people know the correct place to invent the myth of a land-bridge?

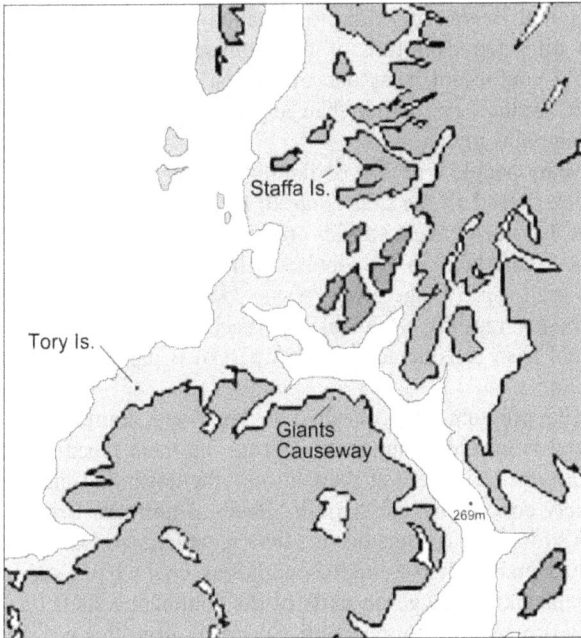

Figure 5.2 The Giant's Causeway
The contours of the Irish Sea floor between the north-west of
Ireland and Scotland based loosely on the 50m depth contour.
This ridge is the terminal moraine of an ancient ice-sheet and
is probably greatly eroded from its original height.

One may recall again that legend preserved by Procopius,
who tells us that the people of ancient times built a wall to
separate the habitable eastern regions from the less favourable
regions west of it. Could this perhaps recall the building of a
protective sea-wall on the low-lying ridge, at a time when it
was above the sea? We may envisage a tenuous north-south
sand-bar something like the modern Chesil Beach, or Spurn
Head, which was at risk of washing-away. The conventional
explanation is that Procopius was making a muddled reference
to Hadrian's Wall, but the allusion is clearly to an ancient
north-south barrier, not to an east-west wall. The specific
references to an era of mild climate would date it to the period
of the mythological Otherworld, as that era was recalled by the

aboriginal Britons of Elmet or Strathclyde during those dark-ages after the departure of the Romans. It may be viewed, should you accept it, as one of our earliest written sources of Celtic mythology; earlier than any Irish or Welsh manuscript that contains myths.

Many people will also be familiar with the legend that Saint Patrick chased all the snakes out of Ireland. Bede also repeats a myth that must have a similar origin to that of Procopius: that if carried there by ship then snakes will die as soon as they scent the Irish air![20] Therefore, says Gerald of Wales, since poisonous reptiles were able to reach the Isle of Man, the ancient Irish considered it to be a part of Britain rather than of Ireland.[21]

The problem of how all the plants and animals reached Ireland is an old conundrum. During the most recent ice-age, when sea levels were at their lowest, the north of Ireland was entirely covered by the same ice-sheet as northern Britain and most of its unique Pleistocene fauna went extinct. Only species able to endure harsh tundra conditions could have survived, perhaps taking refuge on parts of the continental shelf that are today drowned far to the south. Survivors from this era are the unique Irish hare along with pine martens, stoats and otters. DNA research will surely hold the answer to all these questions one day.

Invasive species could only have got in as the ice receded across whatever land link may have existed in the north. In Britain, Grass snakes are rare north of the English Midlands but the adder, though rare, is also found in Scotland. Ireland has only 26 mammal species compared to around 100 for Britain, but the only reptile is the Viparious or 'common' lizard, found also in Scotland and throughout northern Europe. Therefore it was only those hardy species that reached Scotland which could have arrived in Ireland via a northern route.

A survey by researchers at the School of Natural Sciences in Dublin concluded that the most recent date that a land link between Ireland and Britain could have existed was around 16000 BP near the peak of ice age conditions.[22] On this hypothesis Ireland would have been isolated long before the final date that the North Sea link with Britain and continental

Figure 5.3 The Neolithic sea levels around Britain and Ireland based on a pole-shift model. Only a relative-tilt of the sea-floor will give a reconstruction that fits with the description in the legendary sources, which require a submerged 'level plain' in the Irish Sea. In such a scenario the submergence around Cornwall and the Irish coasts would have been much deeper.

Europe was broken. However, on reading the study, one may see that there is no new evidence; no one has actually sampled the sea-bed; it is based on glacio-eustatic modelling again and assumes a parallel rise of world sea level as the northern ice cap is believed to have melted. The concept of a pole-shift is of course not given a thought. However, if the glacio-eustatic theory of sea-level change were shown to be flawed then any modelling based upon it would be similarly invalidated.

It is perhaps understandable that the Irish would not want any link to Britain – not even an ancient one. One may put back to the authors the fundamental question that they posed at the start of their analysis and did not answer. If there was no Holocene land bridge then how did the animals get to Ireland?

Some Surprising Coincidences?

The trail of evidence has narrowed down where and when a catastrophic submergence in the Atlantic could have occurred. This pattern points to the Britain, Ireland and western France in the years just before 3000 BC – the Middle Neolithic era. The empire that supposedly existed before the submergence was remembered into historical times by the inhabitants of a 'central island', which could only refer to Anglesey, or perhaps Mann. We find there the sacred island of the Druids in that place to the west of Britain to which Gaulish tradition believed that the souls of the dead were ferried, and apparently in the same region that Mediterranean myths would suggest was the location of the Underworld.

We also find within Irish myths a belief in a happy Otherworld existing in the form of a great plain beneath the Irish Sea in the vicinity of the Isle of Man. From other Irish sources we find tales that remember this land as a real place that formerly existed and was submerged by a sudden overflow of the sea. We also find local legends that recall the time when Scotland and Ireland were linked by a land bridge; together with legends of the submergence of a city or fields around British coasts. The physical evidence of submerged forests and peat deposits around the costs would suggest the late fourth millennium BC as the likely era when these events occurred.

So far this is only half the evidence, for we have not yet looked at the British myths, or the archaeology, or the data from the coasts and sea bed themselves.

Notes and References

[1] Caesar, The Gallic War, V, 12-14
[2] Strabo, Geography, 3.3.5
[3] Strabo, Geography, 4.4.5
[4] I have investigated Druid astronomy more fully in *Under Ancient Skies*.
[5] Daily Post Wales, 12 December 2017
[6] Pliny, Natural History, XVI,250
[7] Diodorus Siculus, III.33.8-9
[8] Herodotus, II, 51-2
[9] See also Diodorus Siculus citing Hecataeus: Book II, 47,4-6; also III.64
[10] Lucan, Pharsalia, I,450-460; from the translation of A.S. Kline (2014).

[11] See here, Chapter 6 regarding the Nemedians' raid on the golden tower.

[12] The term 'faery' should imply the ancient real people underlying a story, rather than a 'fairy' as an invented supernatural being in a fiction.

[13] For example, in a late story: *The Destruction of Da Derga's Hostel*, we are told of three Fomorian giants, each with three heads; and each one could eat a salted pig all to himself!

[14] Peate Cross T. & Harris Slover C (eds) *Ancient Irish Tales*, George Harrap & Co, London, 1935.

[15] Dr Scott Timpany, reported in The Scotsman newspaper, 25 January 2018.

[16] I shall not follow this thread here; see also William Camden's *Britannia* of 1585, now variously available online.

[17] Or by extension: 'the stepping stones of the under-sea people'.

[18] This is the translation of Gwynn and Thomas Jones, 1974.

[19] In Chapter 14 of *Atlantis of the West* I gave three different translations of this passage all showing similar meaning.

[20] Bede, History of the English Church & People, I.I

[21] Gerald of Wales, History & Topography of Ireland, 48

[22] Edwards, R.J., Brooks, A.J. *The Island of Ireland: Drowning the Myth of an Irish Land-Bridge?* Mind the Gap: Postglacial Colonisation of Ireland. Special Supplement to The Irish Naturalists' Journal., 2008, 19 - 34

6

Coincidences and Contradictions

On the Penwith peninsula of Cornwall we find a concentration of Neolithic monuments, of various ages in such a small area; sites such as Chun Quoit near St Just – with its outsized capstone set on top of large upright stones. Clustered nearby are several similar examples; for example Zennor Quoit, somewhat ruined; Lanyon Quoit, collapsed and rebuilt as we see it. They are very similar to the cromlechs or 'stone-tables' found in Brittany. These monuments, classed as *chambered tombs* or *portal dolmens* by the archaeologists, are all loosely dated between 4000-3000 BC and mark the earliest phase of occupation in this area. Such projects must have required considerable human labour to erect them – but where are the homes and settlements of all these earliest people?

Fig 6.1 Lanyon Quoit – an example of a Middle-Neolithic portal dolmen, which would originally have been covered by a mound of stones.

According to the local folklore; and as made famous in the Arthurian literature, a sizeable community once lived in a lost land called Lyonesse now submerged offshore. We are told that there were formerly many parishes and churches lost between Land's End and Scilly. Even so, it is difficult to believe that quite so many settlements could have been submerged beneath the sea; and so perhaps it refers to the occupation of a much wider inundated region off the north of Cornwall and the Bristol Channel. Shall we believe the folklore, or not?

Archaeologists' theories about the past are based upon classification of physical artefacts in which legend has no place. The non-specialist is presented with an array of often tedious terminology; with classifications of artefacts and monuments that may obscure the underlying truth of what the living people were doing. We may also have their myths and legends – degraded history – but to put the two together is never easy. To add to these we have the skills of the linguist, who can trace the evolution of culture in a region by the development of the prevailing languages; and in the twenty-first century we now have DNA analysis to trace ancestry within the living population. All of these methods must converge on the same story if we are ever to arrive at a true picture of the ancient past; but did you ever hear of an archaeologist revising their theories because they disagree with the folklore; or because they fail to agree with some ancient historian? The ancient bard who was closer to the events has no living advocate in a modern court.

So let us briefly summarise what current archaeology tells us about Atlantic Europe and the Britain and Ireland of five thousand years ago; an unimaginably ancient date that is barely historical in Egypt, but for Europe a complete void.

The Big Stones
Around the Atlantic coasts of Europe, from around 5000 BC onwards we find the Megalithic monuments; a name meaning simply 'big stones' surviving from the earliest archaeology. Before the introduction of radiocarbon and tree ring dating in the 1960's archaeologists supposed that knowledge of farming and settled culture had 'diffused' gradually from the civilized east; and that the Megalithic culture along the Atlantic coasts was inspired by contact with Mycenaean Greece and the near-east. These older 'floating' chronologies were based on cross-dating of pottery and artefact styles linking back to the Egyptian king list, but all had to be abandoned when sites such as Stonehenge and Avebury were proven to be older than the pyramids. That which was a science of pottery has become a science of carbon-14 dating and archaeologists are now as happy to uncover a layer of charcoal as a pottery shard.

In any endeavour to link the legendary past to the evidence in the ground, the historian or mythographer must do the opposite of what comes naturally: instead of seeking the oldest reliable source of information, it is instead the latest discoveries and opinions that have to be cited. Often these must be supported by reports of early excavations that use assumptions and classifications dating from the pre-radiocarbon era. As with Egyptology and climate research the entrenched theories collide with any new suggestion. Therefore, to summarise the currently accepted timescale for the transition from Mesolithic (middle stone age) hunter gatherers to Neolithic (new stone age) farming communities in Britain and Ireland, we may loosely state the chronology that is in current parlance.

Mesolithic	before 4000 BC
Early Neolithic	4000 BC to 3500 BC
Middle Neolithic	3500 BC to 2900 BC
Late Neolithic	3000 BC to 2500 BC
Bronze Age	from 2500 BC

These cultural boundaries are somewhat overlapping and evolve with new discoveries; consensus has changed markedly even since this author's earliest cross-disciplinary investigations. We may see that it is the cultures of the Middle Neolithic that were contemporary with the earliest period of the Egyptian state and with the earliest recorded history.

The transition from Mesolithic to Neolithic corresponds to the clearance of forests for agriculture, but the distinction between the Early and Middle Neolithic is less clear. The archaeologists must make their case for it. However, the transition from Middle to Late Neolithic is sharply defined by changes in monument and pottery styles; and it is contemporary with the transitions of climate and sea level towards the end of the fourth millennium BC. It corresponds to the transition from the mid-Holocene warm period of the Atlantic pollen zone, to the cooler Sub-Boreal zone. The development of farming and the building of the earliest Neolithic monuments around Atlantic coats of Europe coincide with the equable mid-Holocene climate.

Archaeological syntheses will give that the building of megalithic tombs spread from Iberia from around 4500 BC onwards – but it is difficult to separate how much of this a hang-over from those older ideas of diffusion. Around the Atlantic coasts and river estuaries of Spain and Portugal we find chambered tombs that are the oldest known anywhere outside of Brittany. The earliest dolmens or *antas* as they are called by the Portuguese are found to the south and east of Lisbon. Some examples have even had later Christian chapels constructed around them. The distribution continues along the north coast of Spain; and further inland along the Pyrenees we find recognisably similar monuments as far as the coast of Catalonia and on into southern France.

The only gap in the Atlantic distribution is the Gascony coast. Further north along the French Atlantic coast, especially at Carnac in the Morbihan region, we find the complex of chambered cairns and menhirs of which the tomb at Kercado is a prime example. Located within a mile of the Carnac alignments, excavations in the 1970s gave dates for occupation as early as 4700 BC. A later menhir (one of over 3,000 in the region) was positioned on top of it and a later stone circle surrounds it – but it seems the builders respected the presence of the earlier monument. The village of Carnac may be claimed as one of the oldest continuously occupied villages anywhere in the world. As elsewhere in the megalithic region, little trace is found to tell us where the builders of the Carnac complex actually lived.

In Britain and Ireland a similar picture of monument evolution may be discerned, as will be discussed in more detail below. Elsewhere in northern Europe the contemporary early/middle Neolithic culture of megalithic tombs are less widespread but the correspondences of style are evident. In Denmark there are around 2,500 known megaliths, including dolmens and other mounds – thought to be just a tenth of those that formerly existed. Denmark has always been a crossroads for trade between the Atlantic coast and the Baltic Sea. Construction is dated from 3800 BC onwards and although none of the monuments is spectacular compared with Ireland or France they are certainly numerous.

Within the Mediterranean we find even more impressive evidence of the megalith builders and the earlier pre-radiocarbon theories would view these as the origin of the culture. On Sardinia, Corsica and Sicily, the Neolithic transition is now placed relatively early around 6000 BC and the islands were occupied by almost every subsequent maritime culture that dominated the Mediterranean. Here too are found dolmens dating from the Megalithic era.

On the islands of Malta and Gozo we find the temples of Ggantija, Tarxien and others, evolving in use from the earliest occupation around 5000 BC. Again, older diffusionist theories had sought to link these temples to influence from Mycenaean Greece. In the 1950s archaeologist Stuart Piggot would compare these temples to the Court Cairns of Ireland; their crescent-shaped walls were considered to be the inspiration for the more 'primitive' copies in Ireland. Surely, according to the diffusionists, the civilisations of the eastern Mediterranean were the inspiration for the western Neolithic culture? But the revolution of carbon-14 and tree ring calibration rendered these theories untenable.

The temple building period of Malta dates from as early as 3600 BC for Ggantija; and the height of the island culture was contemporary with the Middle Neolithic of Atlantic Europe. The temple at Tarxien with its spiral carvings is thought to be later, from 3100 BC onwards. The temples of Malta are older than the pyramids and predate the first dynasty of Egypt, with monuments far more impressive than anything that survives in the Nile valley from this time. When these cultures flourished, the Sahara to the south was still a grassy savannah and the temple at Saïs in the Nile delta was yet to be built. These matters we have discussed in earlier chapters.

Although the old notion that culture diffused from the Mediterranean to Megalithic Europe may be discredited, the similarities in monument style of which the earlier archaeologists would remark, remain just as valid. However, the evidence from the mythology and the ancient written sources would suggest that the cultural, and particularly the religious influence, was actually in the opposite direction; and in an era much earlier than the diffusionists believed.

The Neolithic Dark Age

Following the forced revision of chronology in the 1970s it became apparent, in British archaeology at least, that there was a dearth of recorded finds from the two-hundred-year period around 3000 BC.[1] To appreciate the significance of this it is first necessary to summarise the evidence of settlement and farming in the earlier millennia.

The introduction of farming along the Atlantic coast is an indicator of the transition to a settled culture from a former hunting and gathering lifestyle. Population growth – reflected in the number of monuments and artefacts that survive – comes only with the transition to settled farming. Climate researchers sometimes refer to the *landnam* – a Danish term used to describe the permanent clearance of the forests as human activity increased. Some of the earliest dates for farming based on wheat are actually found in Ireland from 4700 BC onwards. This acceleration of farming activity corresponds with the elm decline and the other climate changes discussed in chapter 4. It was followed by the period of climate fluctuation known as the Piora oscillation when glaciers in the Alps readvanced – an indicator of colder winters. Whatever happened around this time, it led to a decline of population and abandonment of the fields throughout Ireland and Britain.

In Co Mayo in the west of Ireland a system of Neolithic fields known as the Ceide Fields has been excavated from beneath the accumulated peat. In some places the stone walls run close to Neolithic tombs known as *court cairns*. The cairn at Behy was in use between 3700 and 3300 BC; and the field system is generally dated to c.3500 BC.[2] By inference, other 'Celtic' field systems in Ireland may also be of this age, but cannot be so securely dated. The presence of stone field walls would suggest a society of pastoral farming with cattle rearing and use of woodworking.

Other pollen evidence from the Avebury area of England indicates that between 3100 and 2850 BC the formerly cultivated land was allowed to revert to grassy meadow. In the Lancashire Pennine hills, where the forests had been cleared, peat deposits began to grow as the cooler Sub-Boreal climate set in. This abandonment cannot simply be due to loss of fertility of the soils attributable to unsustainable agriculture, as this should have occurred equally at all eras. It suggests a dual-cause of climate change combined with a real reduction in the number of people needing the land – either they moved somewhere else, or they died. So what did happen around 3100 BC?

This roughly two-hundred-year period of decline would also fit well with the two-hundred-year period that is recalled in the Irish Book of Invasions when Ireland was described as deserted; this falls between the early invasion of the Nemedians and the later colonists. There is really no other era in the physical record that would correspond with the narrative – unless you prefer to ignore the legends and just dismiss the evidence of the Irish stories as 'myths' after all. The Irish and Welsh legends may be giving us a glimpse of what the real people were doing; something that we will never get solely from the hard evidence in the ground.

An Irish Sea Culture Province?

In 1970s the Irish archaeologist Michael Herity presented the idea of an Irish Sea culture province during the Neolithic, going against the then-prevailing current of diffusion from the east. Actually, the idea has older roots. In 1954 Professor Piggot, in his then-influential synthesis, had pointed to the clustering of the older Neolithic Court Cairns around the northern Irish Sea coasts. Herity saw this as evidence of a maritime culture that had spread around the shores of the Irish Sea and its estuaries; and had even traded further afield to Cornwall and Europe. He pointed to the natural barriers to travel along the west of England and southern Scotland before the era of modern roads and bridges; and that coastal navigation would be the open highway for trade and cultural exchange. However, these advances in archaeological thinking proceeded with little comment on the contemporary sea-level and climate changes that must have been experienced by these same people. The builders of the Court Cairns lived through the mid-Holocene period of mild climate and they experienced all the changes of sea level around that time.

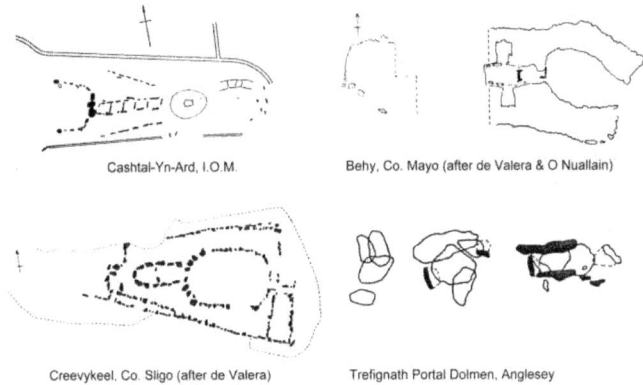

Fig. 6.2 Some examples of Court Cairns and related portal tombs from the Irish Sea region. These are not to scale and reproduced from various sources.

The Court Cairns are grouped around the north and west of Ireland, with a few southern outliers; and over 400 examples have been so far identified. The majority cluster around Carlingford Lough and towards Down and Antrim – an area where the Mesolithic people found abundant chalk and flint for tools and weapons. In structure the tombs consist of larger stones forming a roofed east-west oriented burial chamber, often divided into a number of cists. They are classed as *gallery graves* since typically they possessed no entrance passage. The burial cists would be covered-over by a wedge-shaped mound of smaller stones; and all characteristically possess the crescent-shaped or horned entrance court that gives them their other name of 'horned cairns'. Sadly most have had their covering stones robbed-out long ago. A well preserved, but untypical example of the style is that at Behy, County Mayo and another at Cloghanmore, where twin galleries and one of the capstones survive.

An almost identical tomb style is found in south west Scotland and Argyll which led Piggot to give them their former classification of Clyde-Carlingford tombs. More recently the tendency has been to differentiate these again into separate regional groups: Clyde Cairns and Solway Cairns, etc; and one may see an older name: 'giant's graves' applied to these long mounds. The terminology tends to disguise the likelihood that they probably had the same ritual purpose as the Irish cairns and were built by people of a common culture.

A similar chambered-tomb style is also found on the Isle of Man where we find King Orry's Grave – two similar covered chambers back-to-back that are divided by a modern road. The best surviving Manx example is Cashtal Yn Ard near the east coast. On the Isle of Arran and Bute is another prominent group of these 'giant's graves' of a recognizably similar style.

Archaeologists identify outliers of this architecture in northern Britain. The Tŷ Newydd and Bodowyr cairns on Anglesey may be examples of the Irish style, but that at Trefignath on Holy Island most resembles the Court Cairns. The long cairn at Dyfryn Ardudwy near Barmouth is another example; and the Calderstones, now relocated within a Liverpool park, may have been a horned cairn – though sometimes classified as a later chambered tomb. Another distant outlier of the Irish horned style may be the Bridestones near Congleton, although little survives to see. Northern outliers of the horned style are found in Perthshire and the Hebrides.

Figure 6.3 A drawing of how a typical horned Court Cairn might have looked. Most have had their covering stones robbed long ago, whereas earthen barrows elsewhere have survived.

Typical radiocarbon dates for Court Cairns are Middle Neolithic and range between 4000 BC and 3500 BC but some do show continued usage until much later in the Neolithic. An important indicator here is that some of the Irish cairns have field walls running up to them on surfaces that have been revealed by modern peat cutting. This would seem to confirm that these cairns were in use before the fields were abandoned.

The Court Cairns are not found further south. From the same era in south west England we find the trapezoidal shaped Severn-Cotswold Group. These share the entrance forecourt, but lack the exaggerated 'horns'. Elsewhere we find the chambered long barrows such as those at West and East Kennet in the chalk of Wiltshire. These may all be evidence of a similar burial ritual and religion, but the chambers were usually covered by an earthen mound rather than heaped stones and again they lack the crescent-shaped entrance. The distinction may be due to no more than best use of local materials.

Another variant is found in the concentration of megaliths known as portal tombs formerly termed 'dolmens' or 'cromlechs' that are found at the tip of Cornwall. These were typically constructed of larger megalithic stones and catch the eye the more-so, now that they have lost their covering mounds. The style is closer to the cromlechs on the Brittany peninsula and along the French Atlantic coast as far as Iberia. Some had long mounds, some had round, just as in Ireland.

Of a later form and construction are the so-called *passage graves*. As the name suggests these characteristically possess long entrance passages leading to side-chambers. None is thought to be older than about 3100 BC and they belong to the culture that followed the Neolithic dark-age. Their distribution is superficially similar to the court cairns, but also we find the impressive chambered cairns at Maes Howe and Unstan in Orkney, not far from Ness of Brodgar village first excavated in 2003, and which some commentators would now like to consider as the 'capital' of this entire culture region from about 3000 BC onwards. Also within a few miles of this lies the Skara Brae village of similar age. However, the Neolithic houses at Knap O'Howar on remote Papa Westray are much older, dating from between 3700 and 3100 BC. It is interesting that only here on the shoreline do we find the homes of the living people who built the monuments, which we fail to find in the more populated areas to the south.

Further south on Anglesey, the Bryn Celli Ddu mound is another example of a solar-aligned passage grave. In 2017 archaeologists discovered evidence of an extensive ritual area around the mound. At Newgrange and the other Boyne-valley tombs in Ireland, the passages bear spiral and zigzag carvings, of which the meaning may only be speculated. They likely had some significance related to their astronomical alignment. The passage graves were clearly designed to be regularly re-entered so that their alignment could be viewed. Again, we have only the rigidity of archaeologists' terminology to say that these mounds were originally built as 'tombs'; their purpose may have been more like churches, as the focal point of seasonal religious rituals.

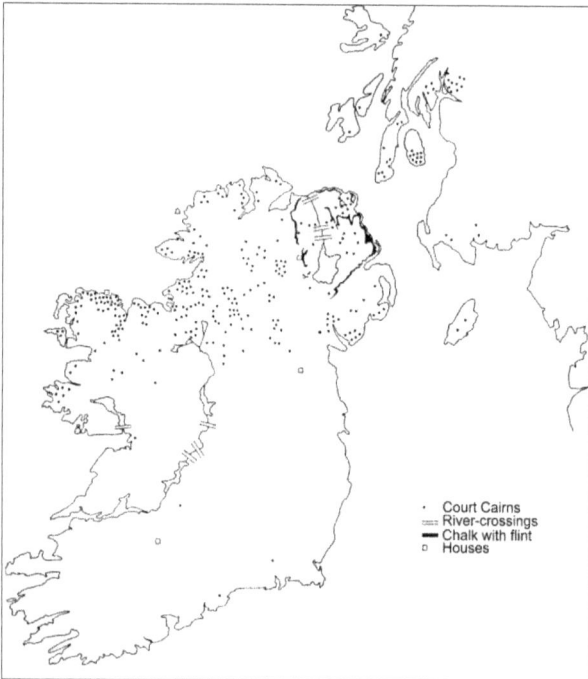

Figure 6.4 The approximate distribution of Court Cairns around the Irish Sea (from Ireland in Prehistory, Michael Herity & George Eogan, Routledge, 1977).

It is these tumuli, close to where the later iron-age capital of Tara was situated, which are usually identified with the 'faery-mounds' of Irish legend. Now superbly reconstructed, the passage grave at Newgrange gave a carbon-14 date around 3150 BC and is aligned on the midwinter sunrise. Other passage-graves nearby at Knowth and Dowth also exhibit evidence of astronomical alignments. This archaeologically rich area contains a number of barrows and standing-stones of various ages as well as a stone circle.

We should just note at this point – as this brief archaeological summary is intended only as a relevant-introduction for the non-specialist reader – that the astronomically aligned stone circles and henges such as Stonehenge and Avebury, that occur throughout Britain and Ireland, are all monuments of the Late Neolithic. They seem to have superseded the chambered tombs in whatever ritual or calendrical purpose they served. We do not find any example that is older than the Neolithic 'dark-age'.

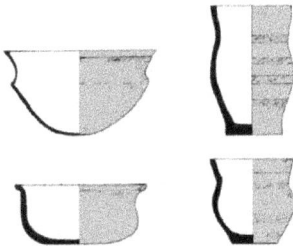

Fig. 6.5 Contrasting pottery styles: typical examples of Irish Sea ware (left) and later Scottish beakers.

Other indications of a common culture during the Middle Neolithic come from the artefacts. A style of pottery known as Irish Sea ceramic ware, sometimes called Lyles Hill or Grimston ware is found all around the Irish Sea, but is not attested later than 3000 BC. It was a variant of a pottery style termed Neolithic 'A' or Western Neolithic, since the early excavations of V. Gordon Childe in the 1930s.

A radical change in pottery style gives the clue to a change of culture or the arrival of new invaders into Atlantic Europe. From around 2900 BC onward we find a transition to individual burials in round barrows rather than chambered tombs; and along with these burials, a new style of pottery – the bell-shaped 'beakers'; and perhaps associated with first known usage of copper. Specialists also remark on a change in human physical type, from the gracile (dolichocephalic) skulls of the older tomb-builders to the round-headed (brachycephalic) skulls found in the later round barrows. Remember the old adage: long skulls in long barrows; round skulls in round barrows!

The entrenched terminology of the archaeologists has led to one of the more infuriating classifications. To refer to these new pottery-makers as 'the beaker people' or 'the beaker culture' tells us almost nothing about them. We might just as well call the Romans: 'the amphora culture' and debate whether there was once a great Mediterranean empire of 'Samian-ware people'. However, more recent evidence from human DNA would suggest that the change of culture during the Late Neolithic does indeed coincide with a wave of immigrants from the east of Europe.

Such then is the gist that we have from archaeology; but it can tell us little about the organisation of the living people. From the (one has to say: dull) reports of the archaeologists one might think that ancient people had nothing better to do during their short lives other than to make pottery and build burial mounds. The true value of archaeology comes when it can offer a chronology from the artefacts to confirm a historical culture, for which we have named kings and events in their reigns. This is how archaeology is employed in Greece and Egypt; but where we have no history to compare then it often fails us. It cannot confirm whether the mound builders had the same language or religion. It cannot tell us whether they were small peaceful communities, or vassal subjects of some distant despot. We may ask: where did they all live, these mound builders? Where are their ruined cities? Where is the Megalithic equivalent of Jericho? Where are the palaces of the Megalithic kings and queens?

To return for a moment to the concept of a Clyde-Carlingford Culture it may be pertinent to quote some of the words of Professor Piggot in his influential pre-radiocarbon synthesis of British archaeology:

> ...the evidence of the grave-goods suggests that the restricted distribution of Lyles Hill ware, of probable Yorkshire derivation, may imply that the makers of this pottery colonized Ireland and probably Man from the north English coast, but avoided western Scotland since it was already occupied by people of a Neolithic culture who are likely to have been the builders of the Clyde-Carlingford tombs in that region.

Of course in the 1950's Piggot was placing this culture region around 2000 BC based on cross-dating.[3] We now know that it falls more than a thousand years earlier. We may also be sure that Piggot never intended his appraisal to sound quite so similar to the legendary invasions of the Nemedians and their wars with the Fomorians that we find in the Irish Book of Invasions. If only we could apply archaeology in this way to authenticate our legendary heritage then just imagine how much more history we might have about our ancestors.

The Irish Sea Coast

In the early 1950s the geomorphologist Ronald Kay Gresswell explored the sea coast around Britain and Ireland and published several books and papers, with a particular interest in the coastline of his native Lancashire.[4] Sadly neglected today, his work would build upon the pioneers of British climate & sea level research such as Harry Godwin, and J.A. Steers; and in the era before radiocarbon dating, it relied principally upon the pollen zone evidence to supply an approximate chronology for the coastal evolution.

From his own surveys of the beaches around the Lancashire and Welsh coasts together with reports from older witnesses Kay Gresswell proposed that a low wave-cut notch in the flat Lancashire sediments represented an early-Holocene beach; this, he believed, marked the furthest inland penetration of the sea following the melting of the former British ice sheet.[5] Clearly this beach had to be older than the tree stumps and peat deposits that sat on top of it, which had been exposed from time to time on the beaches between Southport and the Wirral. These oak and birch forests he assumed (in the pre-radiocarbon era) to be contemporary with the Atlantic pollen zone at the peak of mid-Holocene warmth; and he assigned the raised beach inland to the earlier Boreal period.

The consensus in the pre-radiocarbon era was that coastal forest deposits around Britain and Ireland were all created during a single 'submerged forest period' and older research papers speak freely of this. However as more deposits from various ages were discovered this began to be questioned – but it merely suggests that there has been more than one episode of whatever event causes these rapid inundations – it could occur again. More recent radiocarbon dating of the submerged coastal forests confirms that Kay Gresswell was at least partially correct in his chronology. The deposits at Morecambe Bay, South Lancashire, Cardigan Bay and Bridgewater all point to a marine transgression around 3200 BC corresponding to the Middle Neolithic period of human development.[6]

The hypothesis would suggest that during the mid-Holocene warm period the sea had retreated far out from the present coast for a period sufficiently protracted for mature forests to grow. Since the Manx-Furness ridge is shallow, it therefore followed that this ridge would have been exposed and so this must have been the era at which animals were able to reach the Isle of Man. This would require a lowering of the sea, during a warm era, by at least 30-40 metres from present coasts.

So why did this hypothesis of a submerged forest era fail to gain acceptance? The answer is that it simply went out of fashion. The early 1960s was the era when the Milankovich theory of ice-ages came to the fore and the glacio-eustatic theory took control of all ideas about the causes of world-wide sea-level change. The 1960's was also the era when other pioneers such as Rhodes Fairbridge were looking for (and failing to find) a worldwide sea-level curve to confirm the glacio-eustatic theory.

Therefore to geologists and geographers of the 1960s, who had determined that the North Sea was flooded by 8000 BC consequent upon the relentless melting of polar ice, there was simply no mechanism to explain how the Irish Sea could have remained dry at the same era. How could the sea level have fallen (or the land risen) by as much as 30-50 metres at a time when climate was at its mid-Holocene peak of warmth? There was no mini-ice-age to withdraw water from the oceans. In any case, this should have had the same effect in the North Sea and on other coasts worldwide. The fact that the submerged forests do exist shows that, despite the straightjacket of glacial-eustasy, the coastline around western Britain *must* have stood lower during early Neolithic times – and it did so for hundreds of years – to allow time for mature forests to grow. This detail could be quietly glossed-over as an anomaly. Research into Holocene sea-level change has therefore progressed in terms of metre-scale oscillations around current shorelines and no-one has really looked for evidence further away from the coast.

What Lies Beneath?

The eastern Irish Sea is remarkably shallow. East of a line drawn loosely between Anglesey and the Isle of Man and north to Galloway the sea-bed never exceeds fifty metres depth; gently shelving towards the Lancashire coast and the long beaches around Blackpool, Southport and other Victorian seaside resorts. On navigation charts the former channels of rivers can be discerned where they once meandered across this formerly glaciated plain.

To the west of this the continental shelf drops away to between 50 and 135 metres; though still not deep compared to the open ocean, deep enough to convince geomorphologists that there has been no land bridge to Ireland since the Ice Age. Continuing north the sea-bed drops to around 260 metres at its deepest in Beaufort's Dyke, near the point where the ferries cross between Stranraer and Larne; and becoming less deep again off the Antrim coast until we reach the shallowest region between Donegal and the Scottish islands. This is where we encounter the local legends of a land bridge between Scotland and Ireland.

In the eastern Irish Sea we may trace the former river channels where the extension of the rivers that flow from Wales and north-west England must formerly have flowed. At the end of the Ice Age, on any prevailing theory, there must have been a brief period when this was a dry plain above sea level – but geomorphologists assume that it was at best a region of frozen tundra conditions. One may discern the path where the rivers Mersey, Dee and Ribble would have united into a broader river flowing parallel to the North Wales coast, into which the Conwy River, the Menai and lesser streams out of Snowdonia would have flowed to join it in a broad estuary to the north of what is now Anglesey.

A pronounced shallow ridge called the Manx-Furness Ridge runs between Cumbria and the Isle of Man, which Kay

Gresswell argued had formerly linked the island to the mainland. To the north of this ridge the rivers out of Cumbria and the Vale of Eden would have flowed west forming another estuary between Galloway and the Isle of Man.

Between the Isle of Man and the submerged river channel off the North Wales coast is a flat, loosely-rectangular area called the Manx-Furness Basin. We may trace where the former extension of the River Lune must have flowed out of Morecambe Bay towards this plain via the Lune Deep, to be lost in a meandering channel or a braided river across it. To the west of this rectangular basin lies another shallower region, where on modern charts the sea floor rises to a minimum depth of 32 metres.

Today it is possible to ascend Mount Snowdon or Snae Fell on the Isle of Man by a comfortable steam train and even relax in a café at the summit. If you are very lucky it may be clear enough to see right across the sea and observe all the coasts and hills surrounding the Irish Sea. We may presume that any Mesolithic or Stone Age Briton brave enough to ascend the hills could have viewed the entire rectangular plain of what is now the eastern Irish Sea and could have observed all the rivers flowing across it. Perhaps they could even have lived down there? The only point at issue is *at what era* the observer would have been able to witness this spectacular sight. Was it only briefly at the close of the Ice Age; or could it have persisted much later?

In *Atlantis of the West* it was proposed that this rectangular plain and its rivers may have risen above sea-level during the Neolithic era, between approximately 5500 BC and 3100 BC until they were again submerged. This is also the trail of evidence followed here and it reflects the older climate and sea-level research rather than current consensus. The myths and legends that have guided us to this point would all suggest that the submergence of this low-lying region was very rapid. The Welsh and Irish legends preserve a memory of a 'bursting' or overflow of the sea; and we have the physical evidence of the submerged forests around the modern shoreline to suggest that regions which were formerly inland rapidly became an intertidal zone.

All of the myths and legends that we have examined here and in previous chapters concur that the submerged province was formerly 'a level plain of surpassing beauty'.[7] The Atlantis myth and the Egyptian myths of the Elysian Fields offer the additional detail that the plain was rectangular; other sources are less precise. From Irish myths we are presented with a view of the Otherworld as a submerged flowery plain in the vicinity of the Isle of Man. All of this contradicts the conventional theory offered by the geologists. So should we believe the legends, or not?

It was also proposed in *Atlantis of the West* that this inundation was caused by a pole-shift that occurred around 3100 BC. Although such an overflow would have been sudden and unexpected, the consequent wobble of the axis (the Chandler Wobble) and its associated pole-tides would have left the region as a muddy intertidal zone for around twenty years, followed by a longer phase in which coastlines worldwide were settling and adjusting to the new geoid. This is the period when the coastal forests were drowned and the new shoreline became stable. In the millennia that followed, normal coastal erosion and deposition processes have built the cliffs and beaches that we see today.

A rapid submergence of 30-50m at the latitude and longitude of Britain would require a pole shift in the order of a quarter to one third of a degree of latitude along the line of maximum effect. Remember the four quarter-spheres pattern discussed in chapter 4. This rudimentary estimate is achieved by nothing more complex than ellipse geometry; the difference in radius of the oblate Earth at this latitude, assuming a line of neutrality running from Scandinavia down through Italy and North Africa; submergence to the west; emergence to the east; with the maximum effect along a line drawn from Ellesmere island down through America. Adjustment of the geoid elsewhere would have taken place in the oceanic crust and at the mid-ocean ridges.

It is not possible to derive this picture of a Neolithic emergence and submergence in the eastern Irish Sea basin solely from the glacio-eustatic theory of sea level change. The Manx-Furness basin is inclined between -50M at the longitude of Anglesey to zero at the Lancashire coast. It is only feasible to make this a level plain, as is described in the various legends, by a relative tilt of the crust which has altered the local geodesy.

What are the alternatives to a radical pole-shift hypothesis? A conventional 'earthquake-tsunami' explanation, to incline and submerge such a large shelf of land, would demand a quake that exceeds any known magnitude; it should have left obvious geological evidence around the coasts. Recall again the Lisbon earthquake and the Storegga slide. Could rebound after ice melt have unlocked the dormant fault lines that run beneath the Irish Sea and Scotland? There is no known evidence of such a mega-quake or tsunami, in either the physical geography or in the extant legends; and no conventional mechanism could explain such a quake in a tectonically stable region. In any case, such a huge subsidence of land would itself have changed the axis of figure – its rotational balance; it would itself have triggered a wobble of the axis and a pole tide.

Although the modern Irish Sea is a region of oil and gas exploration, there has to date been very little examination of the sub-surface geomorphology. Such investigation as there is always starts from the conventional assumptions that the entire area formerly lay beneath an ice sheet and that tundra conditions prevailed until the Irish Sea was submerged by the rising sea at the same time as the flooding of the North Sea. Any features found on the floor of the Irish Sea are therefore usually assumed to be of periglacial origin.

In 1987 Dr Robin Wingfield of the British Geological Survey published side scan sonar sweeps of the Irish Sea floor, which were summarised in an appendix to *Atlantis of the West*. The sonar sweeps were taken NNW of Holyhead and Holy Island, together with another taken off the Irish coast. He identified 'patterned ground' described as 'large sorted polygons' and 'nets'. These he likened to typical features found in permafrost after ice melting. A larger feature also identified as of periglacial origin, was a 75m near-circular saucer-shaped feature, which he suggested was a pingo, a collapsed periglacial feature: a mound that subsided after the supporting

162

ice had melted. Certainly this entire area was formerly glaciated, but it would be useful to see more such sonar sweeps further from the coast.[8]

It should be stressed that Dr Wingfield was an eminent geologist who closely followed the standard theory of glacial geomorphology. In a 1995 paper he proposed a model for the coastal evolution around British seas.[9] The model proposed that around 12,000 years ago the raised beaches around current coasts were formed showing that any tectonic displacement had not had any significant effects on the coastline since that time. Using the conventional glacio-eustatic model he derived that between 12,000 and 9,000 years ago, relative sea levels varied between 80 m above and 160 m below present mean sea level. He would further propose land-bridges between Britain and south-west Ireland and between Britain and the Isle of Man after 11,350 years ago.[10] At that era Britain and Ireland would have looked more like Greenland: cold and uninhabited.

One may see therefore that this conventional geology together with the more recent Irish study summarised in the previous chapter would not support any suggestion of a land bridge between Britain and Ireland during the warm mid-Holocene era. Yet, the submerged forests do exist; and all the legends suggest a warm era, not tundra; and the animals did somehow reach Ireland and the Isle of Man.

The Welsh Myths

We have yet to examine what Welsh mythology and legends have to say about the nature of the Celtic Otherworld. This is preserved in the Four Branches of the Mabinogion and in the Welsh triads; which are found in the so-called Four Ancient Books of Wales.[11] Although we can be sure that the mythology existed well before these medieval manuscripts, there is little confirmation as to what form they took. Therefore, as a potential source of evidence about the ancient past they have a similar status to the Irish mythology, with a thousand years more of unreliable oral transmission compared to Classical mythology. No version of the myths predates Christian influence; and modern interpretation is further subject to scholarly preconceptions as to how much derives from continental Celtic belief and how much of it might be pre-Celtic. However, we are not here to dissect Celtic mythology in detail, rather the nature of the history and geography that underlies it.

Almost no written history survives from the post-Roman Dark Ages and Nennius, writing in the late eighth century, apologetically tells us that he made a heap of such few sources that he could find. He tells us that the Britons had no skill and had set down no written record; a criticism perhaps a little unmerited as the Bardic philosophy was always intended to be preserved orally and its continuance was proscribed by the Romans as part of their suppression of the Druids.

In examining such evidence as the myths may offer us about pre-Roman and Neolithic times, we should treat the Welsh sources rather as *British* legends, for they preserve beliefs that were probably held by all the British tribes before their preservation became concentrated in the Welsh-speaking principality. That part of ancient history which was not destroyed during the Roman conquest, or further distorted by Christian conversion, was tested again by the imposition of the English language.

In Welsh mythology the Otherworld may be recognised under the name *Annwn*. The name means something like 'the lowest place', or 'lowest part'. Although many similarities are evident between the Irish Otherworld and its Welsh equivalent there is one very clear difference. The Welsh heroes do not need to sail there in ships; they can simply walk into it. We realize that the hero has strayed from places that we know and from geography that we understand into an otherworldly landscape of castles and beautiful people, with no discernible boundary.

The mabinogi of *Branwen Daughter of Llyr* was briefly introduced in the previous chapter, in discussion of the former land bridge to Ireland. The gist of the story is that, as so often happens to princesses, Branwen is despatched to Ireland to marry the king. She is described as 'one of the three great queens of the island' and the most beautiful girl in the world! King Bran (a giant who seems to vary in size according to the needs of the story) sets off to Ireland with his fleet of ships to deliver his sister to the wedding. Much later however, Bran hears that Branwen has fallen from grace and demoted to doing cooking and kitchen duties. He sets off with his army to walk and wade around the coast, crossing the two rivers, in order to make war on the Irish king Matholwch and to bring her home. Stripped here of its mythological decoration, we may perhaps glimpse the opposite side of one of those Irish invasion myths, but seen here instead from the 'Fomorian' viewpoint.

Here, the Otherworld seems to recall a real historical place rather than somewhere encountered in a dream. It is as if Bran has been transformed into a giant because the incredulous bard who recorded the tale could not understand how else he could wade across to Ireland. Again, we may suggest that the reference to the two rivers that formerly separated Britain from Ireland was included specifically to set an early date for the story, which in this context would place the events at a time before the submergence of the kingdoms. One should never expect consistency of detail in a myth, which is perhaps a fictional reconstruction of real ancient events.

Other Welsh stories prefer the use of a hero-figure who either dreams of, or wanders into, the Otherworld, giving us little idea of its supposed era. For example in the mabinogi of *Pwyll, Prince of Dyfed* in the First Branch of the Mabinogion, the location of Annwn seems to be adjacent to Dyfed in South Wales, as its king Arawn swaps places with the ruler of Annwn for a whole year.

In *Preiddeu Annwfn* ('The Spoils of Annwn') a medieval poem in the Book of Taliesin, the timeless hero figure of King Arthur is employed to make voyages to otherworldly kingdoms and fortresses within Annwn. Ostensibly, his purpose is to find and retrieve a magical cauldron from an Otherworld palace. The poet gives us various names by which this castle was known, which we may compare with the descriptions of the castles and palaces elsewhere. He describes it as *Caer Siddi*: 'the faery fortress', *Caer Fedwid*: 'fortress of the mead-feast', *Caer Pedryfan*: 'four-cornered fortress', *Caer Rigor*: 'intractable fortress' and *Caer Wydr*: 'the glass fortress'. The repetition of the verse is that three shiploads of men went with Arthur but only seven men returned from the expedition.

In the version of the Irish myth of invasions that is given by Nennius, we again encounter this glass tower, but in his version the castle is seen emerging out of the waves, showing us that it has been overwhelmed. Here we find:

> Three sons of a warrior of Spain [Nemed?] came with thirty keels...Later they saw a glass tower in the midst of the sea, and saw men upon the tower and sought to speak with them but they did not reply...

They return to raid the castle, but as they set foot at the base of the fort the sea returns and the majority are drowned except for a single ship which continues on to Ireland. Clearly it is a variant of the same myth as is recorded in the Taliesin poem, but the Irish hero Nemed substitutes for Arthur. This memory of a 'glass' castle is enigmatic: it may refer to some outlandish external decoration, perhaps with amber.

In the eleventh century Irish Book of Invasions we have a third variant of the same legend. Nemed comes with thirty-four ships to settle in Ireland and on the way they encounter the tower in the midst of the sea, but here the tower is instead described as 'golden'. We are told that the tower was submerged at high tide and revealed when the tide was in ebb. Greedily they seek the gold but all except Nemed's own ship and a few survivors are drowned by the rising tide.

Why is this precise detail about the tidal submergence of a tower included as an unnecessary aside within an invasion narrative? Perhaps it was put there to inform us that the tidal range was formerly much greater than we experience today. The modern tidal range in the Irish Sea is less than five metres, so it would not be much of a tower. Perhaps it is used to date the story to the period just after that same overflow of the sea as is described in the Mabinogi of Branwen, which had drowned the kingdoms? We might despair that the bards have not left us more detail about the submergence of the kingdoms. Perhaps such legends did once exist, but now all that remains is the local folklore about sunken cities lying off Wales and Cornwall. Clearly, the lost golden castle and its cauldron represented a vitally important aspect of Bardic religion and philosophy.

The Dream of Maxen

Welsh scholars may say that *The Dream of Maxen* is not the best of the Welsh stories; certainly its structure seems somewhat anticlimactic to a modern reader. However it is the most useful in what it can reveal about the nature of the Celtic Otherworld.

The hero is Maxen. He is a semi-historical figure whom we glimpse just at that point where the underlying historical personality is being transformed into a legendary character. The story recalls *Magnus Maximus*, that Spaniard who was proclaimed as emperor by his own troops in AD 383 while serving in Britain. However the twelfth century bard who composed the extant version of the tale has partially confused him with the emperor *Maxentius* who actually reigned from AD 306-312. The real Magnus briefly took control of Gaul and Spain, awakening memories of former glories among the Romano-Britons; but he was ultimately defeated and executed by Theodosius in 388. He was remembered by the Welsh as something of a local hero who rose to be emperor, but the truth is that he denuded the British province of its defences. Nennius records that his troops were granted land in Armorica (Brittany) and never returned home. This was the origin of the colony to which many more Britons would later flee to escape the invading Saxons.

The story in essence is that the emperor Maxen falls asleep while out hunting and has a wonderful dream of the Otherworld; and he sees there the most beautiful maiden with whom he becomes besotted. On awaking he becomes certain that she is real and sends out his messengers to search the world and find her; but after a year they return bewildered. Maxen, who is now styled as king of the Romans, goes back to where he was hunting and tells the messengers to head west; and so off they go again and this time they find the otherworldly land, which is revealed as Britain and the regions around Mount Snowdon and Anglesey. They find the palace and the maiden and so return to Maxen, who sets-off with his entire entourage to seek the hand of the maiden in marriage.

The remainder of the story leaves behind the Otherworld. Maxen stays in Britain for seven years with his new love and the Romans make a new emperor; and so he has to gather his army of Britons to win back his empire. We may recognise loosely the career of the real Magnus as he conquers France and goes on to lay siege to Rome and captures it by a clever subterfuge. The soldiers then wish to return home to Britain, but we are told of Kynan and his followers, who remain in Brittany to establish a colony; and that is the origin of the province. Since we possess the true history, we may observe here the process by which lost history decays into legend.

Our focus should fall rather upon the description of the Otherworld that is preserved via the agency of Maxen's dream. Quoted here is part of the nineteenth century translation by Lady Charlotte Guest: her other translations are sometimes criticised by Welsh scholars for accuracy, but her version of the dream still conveys the details for us in a most pleasing and poetic sense:

And he saw a dream. And this is the dream that he saw. He was journeying along the valley of the river towards its source; and he came to the highest mountain in the world. And he thought that the mountain was as high as the sky; and when he came over the mountain, it seemed to him that he went through the fairest and most level regions that man ever yet beheld, on the other side of the mountain. And he saw large and mighty rivers descending from the mountain to the sea, and towards the mouth of the rivers he proceeded. And as he journeyed thus, he came to the mouth of the largest river ever seen. And he beheld a great city at the entrance of the river, and a vast castle in the city, and he saw many high towers of various colours in the castle. And he saw a fleet at the mouth of the river, the largest ever seen. And he saw one ship among the fleet; larger was it by far, and fairer than all the others. Of such part of the ship as he could see above the water, one plank was gilded and the other silvered over. He saw a bridge of bone of the whale from the ship to the land, and he thought he went along the bridge and came into the ship. And a sail was hoisted on the ship and along the sea and the ocean was it borne. Then it seemed that he came to the fairest island in the whole world, and he traversed the island from sea to sea, even to the furthest shore of the island. Valleys he saw, and steeps, and rocks of wondrous height, and rugged precipices. Never yet saw he the like. And thence he beheld an island in the sea, facing this rugged land. And between him and this island was a country of which the plain was as large as the sea, the mountain as vast as the wood. And from the mountain he saw a river that flowed through the land and fell into the sea. And at the mouth of the river he beheld a castle, the fairest that man ever saw, and the gate of the castle was open, and he went into the castle.

And in the castle he saw a fair hall, of which the roof seemed to be all gold, the walls of the hall seemed to be entirely of glittering precious gems, the doors all seemed to be of gold. Golden seats he saw in the hall, and silver tables. And on a seat opposite to him, he beheld two auburn-haired youths playing at chess. He saw a silver board for the chess, and golden pieces thereon. The garments of the youths were of jet black satin, and chaplets of ruddy gold bound their hair, whereupon were sparkling jewels of great price, rubies, and gems, alternately with imperial stones. Buskins of new cordovan leather on their feet fastened by slides of red gold.

And beside a pillar in the hall, he saw a hoary-headed man, in a chair of ivory, with the figures of two eagles of ruddy gold thereon. Bracelets of gold were upon his arms, and many rings were on his hands, and a golden torque about his neck; and his hair was bound with a golden diadem. He was of powerful aspect. A chess-board of gold was before him, and a rod of gold, and a steel file in his hand. And he was carving out chess-men.

And he saw a maiden sitting before him in a chair of ruddy gold. Not more easy than to gaze upon the sun when brightest, was it to look upon her by reason of her beauty. A vest of white silk was upon the maiden, with clasps of red gold at the breast; and a surcoat of gold tissue upon her, and a frontlet of red gold upon her head, and rubies and gems were in the frontlet, alternating with pearls and imperial stones. And a girdle of ruddy gold was around her. She was the fairest sight that man ever beheld.

One may see similarity in this description not only with the Celtic Otherworld of other Welsh and Irish myths, but also with the description of the city and plain surrounded by mountains that is preserved in the Atlantis myth. In particular we have the passage: 'and he saw many high towers of various colours in the castle' which echoes again the golden-glass tower seen by Nemed; and the description in Plato's Critias of a palace outlandishly decorated with bronze and orichalcum.

One may also liken it to the island and plain in the passage previously quoted from Diodorus; that plain and habitations on the fabled Atlantic island from his North African source. One may also see similarities here with the Elysian plain as it is drawn on Book of the Dead papyri. The inhabitants, as we find in Greek mythology, are similarly described as golden haired and beautiful. Note also the description of the golden torque about the man's neck, just as the Fomorian prince Elotha is described in the Irish invasion myth. Such coincidences cannot arise by different authors, hundreds of miles and hundreds of years apart, each inventing a fiction. They must all stem from the same root memory.

The description of the plain itself is perhaps the most revealing of all: 'between him and this island was a country of which the plain was as large as the sea'. For comparison, an alternative translation gives this as: 'between him and this island he saw a land where the plain was as broad as the sea...'.[12] Which sea could this refer-to? In this context it can only be comparing the dimensions of the Otherworld plain to the dimensions of the Irish Sea.

When the messengers travel to Britain, they eventually find the otherworldly land of Maxen's dream and the geography is explained for us. Again in Charlotte Guest's translation we find:

And thereupon thirteen messengers of the emperor's set forth. And before them they saw a high mountain, which seemed to them to touch the sky... And when they came over this mountain, they beheld vast plains, and large rivers flowing there through. "Behold," said they, "the land which our master saw."

And they went along the mouths of the rivers, until they came to the mighty river which they saw flowing to the sea, and the vast city, and the many-coloured high towers in the castle. They saw the largest fleet in the world, in the harbour of the river, and one ship was larger than any of the others. "Behold again," said they, "the dream that our master saw."

And in the great ship they crossed the sea, and came to the island of Britain. And they traversed the island until they came to Snowdon. "Behold," said they, "the rugged land that our master saw." And they went forward until they saw Anglesey before them, and until they saw Arvon likewise. "Behold," said they, "the land our master saw in his sleep." And they saw Aber Sain, and a castle at the mouth of the river.

At first the messengers see the same otherworldly scene as Maxen, for they recognise the plain and the city as they travel over the unnamed highest mountain. Evidently they have strayed into the regions of the Otherworld. The journey then returns to the real world, but now they have to return in the great ship to arrive again at Britain, and they identify the mountainous land with Snowdon; and the level region becomes Anglesey; and the city becomes one with the Roman fort at Aber Sain (Caernarvon) and then the story proceeds with the semi-historical account of Emperor Maxen and his expedition to Rome.

It is fundamental to appreciate that the Dream of Maxen *is not a story about Maxen!* His character is merely a vehicle; a hero-figure who would be known to the contemporary audience and who replaces some earlier and forgotten hero in a visit to the Otherworld. The true purpose of the narrative is solely to preserve the description of the Otherworld and to keep it alive; it doesn't matter that a later bard has morphed it into a semi-historical yarn about a Roman usurper and has tagged on to the end an ill-fitting account of the emigration of Britons to the continent. The pagan bard who composed the original story knew that to preserve the memory of the Otherworld he must keep it meaningful to his audience. However, to the Christian dark-age bard who committed it to writing, the Otherworld had already become just a place of fairy-tale fiction as it seems to us today.

At first we are given the geography of the Otherworld in the dream and we may also compare how closely that description matches to the submerged topography that lies just off the North Wales coast. We may see there, where broad rivers once flowed across the level regions. The city with its otherworldly castle, we are told, lies at the mouth of the river.

171

We may also see how closely the reconstruction resembles those Mediterranean stories of a paradise island lying out in the Atlantic, as it was described by Diodorus Siculus. You may also see, *if you wish to see it*, the similarity between the submerged topography on the bed of the Irish Sea and those depictions of the Elysian Fields that we find in the Egyptian Book of the Dead, which show the city of the gods lying at south-west of the rectangular plain. It requires imagination to see this pattern and some people will never perceive it. Even if you take the Atlantis legend completely out of such a reconstruction, there remains enough convergence from the Irish and Welsh legends alone to suggest that the most likely place to seek the location of a submerged city and its palace would be on the highest point of the sea bed between Anglesey and the Isle of Man. If it be not there then a site just north of the Anglesey coast would as well fit the trail of evidence – where the submerged river channel along the North Wales coast formerly met a much reduced Irish Sea. Here then, in the Dream of Maxen, we may see why it was that the location of Anglesey was considered so sacred by the pagan Druids.

Coincidence and Convergence

Consider the number of coincidences and convergence of both legends and physical evidence that you have followed to reach this point. By a process of elimination we have narrowed the focus of investigation to a cataclysm occurring in the late fourth millennium BC; and concluded that the only real place that could fit the geographical requirement is Britain and Ireland. How likely is it that, at the focus of that evidence, we should then find a submerged rectangular plain at the centre of Britain and Ireland, together with local myths and legends that actually remember a rapid submergence; and which describe it as it would have looked when it was above the sea during the warm Neolithic era? How much of a coincidence is it that we find submerged forests of the correct age around these same coasts. How much of a coincidence is it that archaeologists find correspondence with the monument styles and pottery all around the Irish Sea coasts, together with indications of a complete change of culture at this period of prehistory?

It is difficult-enough to find lost archaeological sites on land, let alone beneath the sea-floor. If the theory has merit, then the subsurface of the Irish Sea floor should also yield Middle Neolithic artefacts and we should expect to find submerged Court Cairns like those all around the modern coasts. We should expect to find the scattered remains of Neolithic villages not unlike Skara Brae, concealed beneath five thousand years of silt deposition.[13] The astute reader may have already noted that the distribution of the Court Cairns around the Irish Sea coasts and islands matches almost exactly with the region that Irish invasion legends ascribe to the people known as the Fomorians – the survivors of the catastrophe. But it is important in such a synopsis not to stray too far beyond a pure summary of the evidence.

With which link in the chain of evidence do you disagree? If a submerged city and its surrounding plain be not in the suggested place then how do you explain the coincidences? Even if no evidence of human culture were found on the sea-bed it is clear that the ancient Celts *believed* it to be real and they preserved a belief that a city with its towers and surrounding fields were submerged there. It remains probable that Greek and Egyptian myths originate from the same belief in the lost Otherworld that we find in the Celtic myths. Despite the prevailing academic caution that the Celtic Otherworld and the Elysian Fields of the Greeks and Egyptians are not the same place, they clearly are related. We may regard them as different views of the same ancient place separated by five thousand years of independent evolution: from history, through legend, to myth.

Notes and References

[1] I think the first to draw attention to this 'dark age' was author Aubrey Burl in 1979; however the same correspondence was noted independently in various papers and syntheses after tree-ring corrected radiocarbon dates became available from the 1970s onwards.

[2] Archaeological excavations at Behy (E747) Stratigraphic Report, Warren et al, UCD School of Archaeology, Nov. 2009.

[3] Piggot, S. *The Neolithic Cultures of the British Isles*, Cambridge University Press, 1954

[4] Kay Gresswell, R, *Sandy Shores in South Lancashire*, Liverpool Univ. Press 1953.

[5] Ibid, Chapter 2. He termed this feature the *Hill House Coast* after a prominent Lancashire landmark.

[6] Heyworth, A. in *Dendrochronology in Europe* pp279-288 (1978).

[7] Diodorus Siculus, V.19.1

[8] R.T.R. Wingfield. *Giant sand waves and relict periglacial features on the Irish Sea bed west of Anglesey*, Proceedings of the Geologists Association, 98, 4, 400-404.

[9] Robin T. R. Wingfield. *A model of sea-levels in the Irish and Celtic seas during the end-Pleistocene to Holocene transition*; Geological Society, London, Special Publications, 96, 209-242, 1 Jan 1995 https://doi.org/10.1144/GSL.SP.1995.096.01.15

[10] In his summary he would describe the derivation of this model as: *"A simple, geometrically based model is developed of the interaction of: glacio-eustasy, from graphs based on coral-reef studies; glacio-isostatic depression with an annular fore-bulge of equal volume, both contracting through the interval considered; and hydro-isostasy, as an enhancement of the other effects by up to 20%"*

[11] This is the title given to the medieval Welsh sources by William Skene in 1868.

[12] Gantz, Jeffrey (1976) *The Mabinogion*, Penguin Books, Harmondsworth; p 120

[13] If we assume a conservative 1-2 mm per year of silt deposition then anything of archaeological interest could now lie 5-10m beneath the sea bed.

7

A Divergent View of the Past

The medieval cathedral builders knew that in order to add strength to a structure would sometimes require a little further buttressing around the exterior. So before we leave behind the theory that a lost Neolithic city, as remembered in Plato's Atlantis myth, might lie beneath the Irish Sea then it may be worthwhile to pull on just a few more of the interesting threads that have been left hanging.

Neolithic DNA

In 2018 a group of researchers published a genome study of DNA taken from west European human remains dating from the Neolithic, specifically the transition from the Middle Neolithic farmers to the Beaker People of the Late Neolithic. Their findings were no less than astonishing.[1]

From a study of 170 Neolithic and Bronze-Age Europeans, of which 80 were derived from Britain and Ireland, they concluded that over 90% of the Neolithic gene pool was completely replaced by new immigrants within the space of just a few hundred years. The study refers to: 'a nearly complete turnover of the population' during the Late Neolithic[2]. They conclude that whereas for continental Europe the expansion of the Beaker pottery and its users might be attributable to trade, for Britain and Ireland it was certainly demonstrative of an influx of new people.

The study further suggests that the earlier Neolithic farmers originated from a mainly Iberian population that had spread along the Atlantic coast of Europe during the earlier phases of the Neolithic (they remark on the presence of the megalithic monuments in this region). However, the study goes on to propose that the people of the Bell Beaker Culture had an eastern European or 'steppe' ancestry, with the majority of those who settled in Britain being related to people from the lower Rhine.

It is becoming clear that there never was a great invasion of continental Celts to Britain during the Iron Age, as our older text books tell us; and that any later trickle of immigrants from the Celtic and German regions would in any case be difficult to distinguish from the later arrivals. There may indeed have been no major influx of new immigrants until the arrival of the Anglo-Saxons over two thousand years later. However, this technique cannot readily distinguish cultural colonisation by small ruling elites, such as the Roman conquest or the later Normans.

Although, as one might expect, the researchers make no reference to the indigenous legends of the native peoples, none of the above would contradict the picture already available from the 'soft' forms of evidence. It would seem to confirm that the Mid-Neolithic Crisis, which we see in the climate and cultural record, does indicate a real reduction in the population of Britain and Ireland prior to the arrival of the continental immigrants.

The Ancient British Tribes

From various classical sources we may find details of the tribes that occupied Britain in the late Iron Age, just prior to the Roman intrusion into the island.

In the Agricola of Tacitus, we are given a useful summary description of the British population at the time of the Roman conquest of the North.

> ...their physical characteristics vary, and the variation is suggestive. The reddish hair and large limbs of the Caledonians proclaim a German origin; the swarthy faces of the Silures, the tendency of their hair to curl, and the fact that Spain lies opposite, all lead one to believe that Spaniards crossed in ancient times and occupied that part of the country. The peoples nearest to Gaul likewise resemble them...

On the relationship to Gaul he comments:

> In both countries you find the same ritual and religious beliefs. There is no great difference in language...

There seems no good reason to dispute these observations from a historian who was closer to the facts than us, even though his opinions are far from complementary about the natives. The question becomes: how ancient are these regional variations?

From the map of Claudius Ptolemy (second century AD) we have the approximate disposition of the British tribes (see figure 7.1). We cannot determine the precise tribal boundaries, or even be sure of their locations, because the mapping itself is not precise.

In the south-west peninsula we find the Durotriges, who are remembered in the name of Dorset; the Cornovii and the Dobunni are shown in the centre of the island, whence they would migrate to Devon and Cornwall (Dumnonia) in Roman times. Other tribes are known that do not appear at all on Ptolemy's map. The Deceangli for example are identified only from inscriptions at Chester and may be recalled in the name of Anglesey. The north of Wales, Snowdonia and the region of the modern county of Gwynedd, was the preserve of the Ordovices tribe, with the Deceangli to their east. In the south of Wales were the Silures, of whom Tacitus remarked; with the Demetae further west.

Away from the coasts we cannot know how far north any invading tribes from Gaul or Spain might have extended. We do know from Caesar and his earlier sources that some tribes 'in the interior' held traditions that they were native to the island. It was not simply a case of complete assimilation of the older population. We may draw comparison from the later influx of Anglo Saxons, where the earlier Britons were pushed into Wales, Yorkshire and Cumbria and into southern Scotland. In the north of Scotland we may see influences from both east and west coasts, which pressed older tribes into the highland glens. Many of Ptolemy's Caledonian tribes must have been small isolated communities.

Figure 7.1 The Map of Claudius Ptolemy

A reconstruction showing the approximate position of the tribes at the time of the Roman invasion and later migrations. The tribes 'in the interior' whom Caesar and other writers described as autochthonous are most probably those of the north west of England, North Wales and western Scotland. Additional tribal names such as the *Deceangli* in Wales and *Attacotti* ('very old-ones') in Scotland are available from other Roman sources.

From the revolt of the Iceni tribe of Norfolk in AD 60 we have a description of Queen Boudicca as tall, with flaming red hair; We may note her claim to be descended from 'mighty men' and that the British were accustomed to women commanders in war. We cannot be sure that she was typical of her tribe, due to the tendency for intermarriage between royal houses, but it does give us an indicator that she considered herself to be an example of the ancient stock.

The earliest descriptions that we have of the Irish from classical sources are even less complementary than of the Britons. To say that Strabo compared them to Scythian cannibals who were even more savage than the Britons would be to cite him politely – but he does add that his source (presumably he means Pytheas) was untrustworthy.[3] However, for Ireland we also have the timeless ethnic information in the Invasion myths – if only these were accorded the status that is given to classical sources. The native Pictish myths of origin also describe the Picts (Cruithne) as immigrants from Scythia; they took Irish wives and some also settled in Ireland.

Prior to the Anglo-Norman reorganisation, Ireland was divided into the four (formerly five) over-kingdoms, or provinces that we know today; the precise origin of these divisions is lost in time. The tribes on Ptolemy's map do not sit well within these ancient provinces and probably date from the early years of Roman Britain, when they briefly considered an invasion of Ireland. We may note the repetition of some British tribal names across the Irish Sea. For example, the Brigantes who occupied northern England appear again on the east coast of Ireland; and the Venicones and the Irish Venicnii may also be related.

The Roman ethnographic information alone does not help us to identify what the ethnic mix of Britain and Ireland may have been three thousand years earlier. The question is of course relevant because it is only the aboriginal inhabitants who could have preserved the stories of the lost Otherworld and of cities submerged around the coasts; and who could have retained traditions dating back as far as the middle Neolithic era. So perhaps this is a good point to restate the only facts that we know with certainty. Some of the tribes in the interior of

Britain claimed to be autochthonous to the island. For Ireland we are told that *all* of the population believed themselves to be invaders or colonists. The invading tribes had three routes of origin: via Spain, from Gaul and from the Baltic. These tribes shared a common culture centred on Druidism, indicating an earlier religious unity; or that they may formerly have been part of larger unified kingdoms. The three points of origin roughly coincide with the three linguistic divisions: p-Celtic (British), q-Celtic (Gaelic) and Pictish. We now know, from the DNA that less than 10% of the tribal populations represent the original Neolithic farmers. So who were these survivors?

British & Irish DNA

In 2016 the results of a UK-wide survey of regional DNA ancestry were published, followed in 2017 by a similar map of Irish ancestry. Such studies are replete with specialist jargon, as one might expect when specialists publish work primarily for discussion among their colleagues rather than for a general readership. As so often the cross-disciplinary researcher must rely upon such specialist conclusions, for we cannot repeat their work. We may however, use it to challenge the long-held doctrines of other disciplines.

The DNA study took volunteers from mainly rural locations whose four grandparents had all come from within 80 kilometres. The researchers were then able to assign these to seventeen regional groups based on the presence of identifiable unique genes; essentially telling us about the population of the late nineteenth century when people were less mobile. It would be useful perhaps to see similar studies with photographs and personal measurements for such individuals, to see if any obvious correspondences of morphology may be discerned; perhaps this will follow.

The study revealed much genetic diversity in the north and west of Britain. In the south-east however, the ancestry pattern is dominated by genomes that most likely demonstrate the European Anglo-Norman ancestry. This in turn is difficult to distinguish from any Roman and pre-Roman immigration from these same regions. Much larger and tighter sampling would be needed to identify the underlying native ancestry component.

For Ireland, the lineages may be seen to cluster within the former provinces of Munster in the south and Ulster in the north, with another distinct group in Connacht to the west. Unlike Britain, there is no obvious correspondence to Ptolemy's Irish tribes. As one might expect there is much overlap in Leinster and the midlands of Ireland, showing the most likely route for intruders, ancient and more recent, via the east coast. There are also correspondences with the west of Scotland, which is to be expected from the historical migrations between these regions; together with another quite unique grouping along the northwest coast of Ireland and the Scottish islands.

It seems to have come as something of a surprise to the DNA researchers that the regional ancestry should so closely match the tribal and provincial boundaries that we have known for so long. There is an expression in colloquial English, which might adequately categorise such an outcome as: 'a statement of the blindingly obvious'. Set aside for the moment more recent changes to the ethnic mix. Local people and those of us who have long studied the ethnography and archaeology have always known that such regional differences existed in the native population. Visitors from other English-speaking countries such as the USA or Australia, where population is more homogenized, will often remark on their culture shock at the plethora of regional accents in Britain: the Brummies; the Geordies; the Aberdonians. These modern language variations conform to the former regional kingdoms of the Anglo-Saxons, the Welsh and the Picts; these in turn sit on top of the ancient tribal territories that were little altered by the Roman occupation.

The Liverpool football fan on the way to visit neighbouring Manchester United has never needed a symposium of academics to tell them that they were a different lot to their Lancashire neighbours. The cheery Aberdonian has always known that they were altogether dissimilar to the fiery Glaswegian. The Cornishman has always known that different people lived over the boundary in Devon. And Yorkshire folk have never needed to be told that they were distinct from the other English around them – but they always thought it was the

Viking in them. Now we know that these regional differences are much older.

The situation now is that not only may we cite the DNA evidence when considering matters of ancient history and archaeology – but we must cite it. Formerly any researcher who suggested non-Celtic ethnic origins based solely on traditional evidence would find the speculation flag raised and they would be sent from the field by the academic referee. That which should always have been obvious has become official science.

In the early 2000s there was an article in one of the Yorkshire newspapers: that modelling agencies would visit Harrogate in North Yorkshire in search of new talent.[4] Apparently the local girls possessed all the right proportions (slim, long-legs, blond, etc.) that such agencies seek. Anyone who stays for any length of time in a region may notice local differences of temperament, morphology, hair colour, etc. One may be amused by the medieval comments of Gerald of Wales: that Welsh people would constantly change their minds and forget their promises from one day to the next – until you have to regularly work with someone who exactly fits the stereotype.[5] But it would be unreasonable to ascribe to all, the characteristics of a few. These little clues can offer us extra evidence about people in the ancient past and one may sometimes notice such identity traits in the legends and in the reports of ancient observers. The people of the past really were just like us!

A separate study of Welsh DNA has shown that the people of North Wales are genetically distinct from the rest of Britain.[6] Professor Donnelly of Oxford University commented that the Welsh DNA could perhaps be traced back 10,000 years to the post Ice Age population; and that markers showed the ancestry of the people of north and south Wales to be further divergent from each other than either is from some of the groups in England. As might be expected, both regions showed connections with Ireland. Another anomaly was noted in north-east Wales, around Wrexham, where DNA typical of eastern Mediterranean origins was noted. However, the research could not confirm that this was of ancient origin.[7]

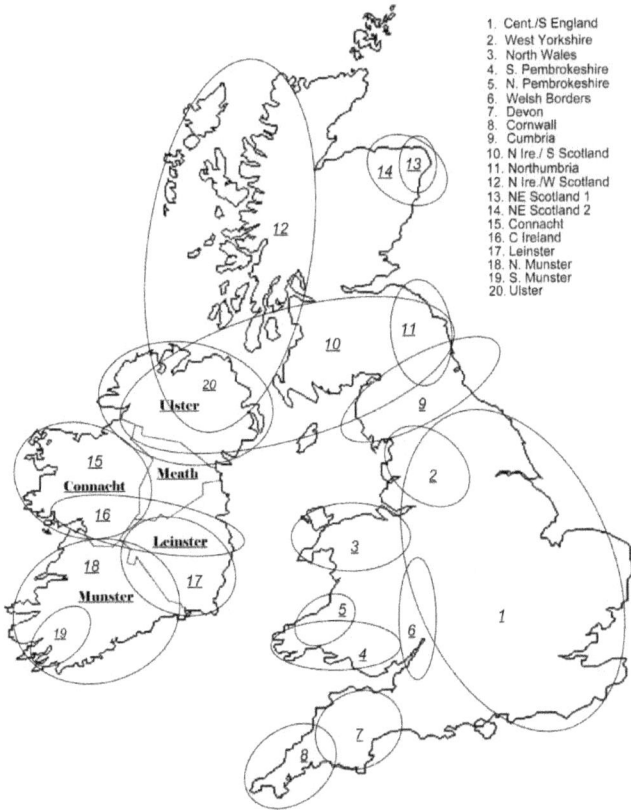

1. Cent./S England
2. West Yorkshire
3. North Wales
4. S. Pembrokeshire
5. N. Pembrokeshire
6. Welsh Borders
7. Devon
8. Cornwall
9. Cumbria
10. N Ire./ S Scotland
11. Northumbria
12. N Ire./W Scotland
13. NE Scotland 1
14. NE Scotland 2
15. Connacht
16. C Ireland
17. Leinster
18. N. Munster
19. S. Munster
20. Ulster

Figure 7.2 British and Irish DNA Clusters

A summary-map showing the most significant DNA provinces of Britain and Ireland based on Nature *Scientific Reports* volume 7, Article number: 17199 (2017). The full study identifies some 30 regional clusters and considerable overlap. The clusters may be seen to conform to the known older tribal groupings and provinces (compare Fig. 7.1), apart from southern and central England where population is more homogenous with continental Europe.

In 2012 the *Yorkshire Post* newspaper published its own survey of Yorkshire DNA. Their analysis based on 200 people of local rural descent, showed some diverse results that were confirmed by the later national study.[8] They concluded that the mitochondrial DNA (inherited only via the female line of descent) from 62% of those sampled were most closely related to people from Iberia and south-west France; and that the female line must have been present in the region since as far back at the earliest post Ice Age population. However the male line shows a different story. The Y-chromosome lineage (which they term 'Pretani') that is prevalent in the population further south was present in only 16% of the Yorkshire males tested. Part of their 'North Sea Germanic' heritage is likely to be of Viking origin. The remainder of the population may date from as far back as Mesolithic times when their ancestors could simply have walked across the Doggerland connection from Denmark.

This unique genetic heritage corresponds closely to the former boundaries of the West Riding of Yorkshire, encompassing the independent Dark-Age kingdoms of Reged, Elmet and Leodis (Leeds), the Yorkshire Dales; and across to the north Lancashire coast. The conclusion is that this was an enclave of older population protected behind the Pennine hills during the period when the invading Beaker Culture took the lowland regions further south and east and had even spread across the sea to Ireland. Beyond the Pennines, another pocket of unique ancestry is found in Cumbria and up to the southern Uplands of Scotland.

It would seem that the females in this older population were preferentially selected over the generations; indicative of a conquered population, where the men are perhaps killed in battle or taken as slaves; or otherwise of low social status; whereas the women are chosen as wives by the dominant males (or perhaps it was the other way round). It may be that the modelling agencies in search of new talent know something that the scientists miss; and perhaps a totally unscientific study of girls' vital statistics by a modelling agency can tell us as much about the reasons for their survival as a map of their genes!

We may now attempt to pursue the statement of Caesar: that some tribes in the interior of Britain, claimed on the strength of oral tradition to be indigenous. The most likely candidates based on the DNA results would be the Brigantes of Yorkshire and the Ordovices of North Wales; and perhaps the tribes of western Scotland and Devon. Aboriginal populations are most likely to survive in locations that are insulated from invaders behind unwelcoming hills and natural barriers, just as we find with more recent populations that were late to be absorbed into the English and Scottish kingdoms. Of course, we may be sure that even as far back as Mesolithic and Neolithic times, tribal differences of origin already existed; and they too will lie somewhere in the ancestry data. At time of writing the study of DNA is only in its infancy and perhaps one day it will be able to tell us precisely where the invaders came from, so that we may again match the evidence to the legends and the archaeology.

Geoffrey of Monmouth and his Myth of Origin

Few experiences could better confirm to us the void that exists in pre-Roman British history than to spend a while trying to make sense of Geoffrey of Monmouth's *History of the Kings of Britain*. Where Nennius would apologise that all he could find was a heap of history – Geoffrey attempts to fill the gaps. Firstly, to declare, that it is not advanced here as a reliable source of history, or even of legends – but it is still a valid place to seek some of those 'fossils' that we may compare with other evidence. Geoffrey claimed that the foundation for his history, dating from 1136, was a very ancient book written in the British language.

In other Welsh manuscripts we do find versions of the *Historia*, with similar pedigree, known as the *Bruts* and it is possible that a lost Welsh version may indeed have been Geoffrey's 'ancient book'. However, most commentators would look no further than the *Historia Brittonum* of Nennius for the source which Geoffrey augmented with miscellaneous Welsh material. Whether it was Geoffrey himself or his Welsh source that filled in the historical gaps, any such hoax must ensure that fiction does not contradict a more respected authority; and that the material will seem familiar to the intended audience. Presumably the author acquired wealth and fame by doing so. As his translator Lewis Thorpe remarked, with Geoffrey of Monmouth 'real history peeps through the cracks' and so one cannot but wonder if the unverifiable material might hold real lost history.

One such example lies in the myth of Origin. We are told that the origins of Britain and indeed its very name are owed to the eponymous invader *Brutus* and his Trojans. Rather like the heroes in those Irish invasion myths Brutus is given Mediterranean origins, but this time Roman rather than Greek. We may wonder if he was a creation of the late Romano-British period, to give the Britons a sense that they were really Romans after all.

The story goes that, returning from the Trojan wars, Aeneas fled to Italy where he fathered Brutus (Nennius calls him 'Britto'). When he grew-up he went to live among his Trojan ancestors and soon rose to be their leader due to his military prowess.[9]   Before setting sail to find a new home for his people, he consults an oracle and the goddess Diana tells him:

> Brutus, beyond the setting of the sun, past the realms of Gaul,
> there lies an island in the sea, once occupied by giants. Now it
> is empty and ready for your people...

Now where has this idea come from: that Britain lay empty and waiting? What about the 'giants'? None of this comes from Nennius. The nearest that we get to it are the legends of the two-hundred years that Ireland was supposedly left vacant. We may suspect that some of the invading Iberian-Greek colonists mentioned in the Irish myths of origin also landed in Britain – but the curious synchronisms of Geoffrey would place all these events after the time of the Trojan wars (c.1100 BC) rather than earlier around 3000 BC.[10] Perhaps it is safest to assume that it is a made-up fiction which has incorporated some well-known mythology already in oral circulation.

Geoffrey offers us an aside from the main history to explore the so-called 'Prophecies of Merlin'. Here we find Merlin, a dark-age wizard or Druid, who gives us a list of portents that were probably never intended to be coherent, but full of double-meanings like a Greek oracle or a modern horoscope. In one place he refers to the 'malice of the planet Saturn' and that the moon will disobey its normal appearances. Saturn we know was a planet that ancient Druids observed with reverence. This gives us a clue that it incorporates a vestige of authentic Druid astrology. The most informative detail however, comes in the enigmatic reference to fluctuating sea-levels:

> The sea over which men sail to Gaul shall be contracted to a narrow channel. A man on any one of the two shores will be audible to a man on the other, and the land mass of the island will grow greater.
> The secrets of the creatures who live under the sea shall be revealed and Gaul will tremble for fear...
> In the twinkling of an eye the seas shall rise-up and the arena of the winds shall be opened once again.

Now once more we may wonder if any of this is based on a grain of truth because, in the normal experience of life, no-one would ever think of such a thing; it is outside our normal understanding of how the world works. Hazards that occur on timescales longer than a human lifetime are forgotten or become myth, just as the residents of Pompeii had forgotten that they lived adjacent to a volcano. The description is a good illustration of what would happen if the sea were subject to a pole-tide. The tidal range would indeed retreat to expose the sea-bed and then, with the period of the pole-tide, return to spill over the land. As in the other examples discussed, there is no mention of deluge or earthquake destruction – the prophecy is here describing an abnormal state of a calm sea.

This experience of an overflow of the sea in some ancient time was remembered along with the local folklore of sunken cities around the coasts; and in the stories about Annwn and the Otherworld. These same beliefs must have been absorbed by the continental Celts further inland from Atlantic shores, when they adopted the religion of the Druids. We find a clue to this in Arrian's history of the campaigns of Alexander the Great. This was the era between 500 BC and 300 BC when the continental Celts briefly expanded their territory, perhaps under a single great-king whose name is lost to us. Around 400 BC they would intrude into Italy, sacking Rome and taking northern Italy from the Etruscans; into Iberia where they took territory from the Carthaginians; and also ranging into Illyria and across the Bosphorus into Asia-minor. Therefore, before Alexander could set-off on his own expedition to conquer the Persian Empire, he needed to secure his western flank from the expansive Celts.

In their negotiations with the Celtic ambassadors the Macedonians sought assurances that they would not break their sacred oath. Alexander asked what the Celts feared most. The reply that they gave to the young king was recorded by his biographer, who tells us that the Celts said they feared no-one, only that the sky might someday fall on their heads.[11] It would seem that the Celts gave to Alexander something similar to the Irish oath:

> If we fulfil not our engagement, may the sky falling upon us crush us, may the earth opening swallow us up, may the sea overflowing its borders drown us...[12]

So we see again here this antique dread that destruction might come from the sky and that the sea could somehow rise-up to envelop them. It must surely have formed a part of the lost Druid teachings. Clearly the Celts viewed such disasters as related and it gives us a clue to why the Druids observed the stars and the calendar so closely, in order to foresee the potential portents of doom.

A similar form of the Irish oath is found in a saga called: The Cattle Raid of Cooley (*Táin Bó Cuilinge*) recorded in the seventh century:

> ...unless the blue-bordered sea come over the expanse of life, we shall not give one inch of ground

And we find the same relict form of words Christianised in a medieval Welsh poem: *The Lament for Llewellyn Ap Gruffudd,* the last independent Welsh king of Gwynedd, which compares his death to a natural disaster of dire proportions:

> See you not that the stars have fallen?
> Have you no belief in God, foolish men?
> See you not that the world is ending?
> Ah God, that the sea would cover the land

We may therefore see further corroboration that the warnings in Merlin's prophecies must predate Christian influence and have passed through from some authentic Welsh traditions.

It would seem unlikely that a Greek biographer would have known of an Irish saga, or any other Celtic source of the myth. The fact that we find this primal fear independently recorded in such divergent sources is confirmation that it was a genuinely ancient terror based on folk memory of a real event.

Herodotus and the Tyrrhenians

Herodotus, who is the oldest historian whose work has survived in its entirety from antiquity, is often treated with scorn and derision for some of his statements, even by other classical writers. Father of History – Father of Lies is his epitaph. Plutarch savagely criticised his accuracy and methods. The Athenian historian Thucydides dismissed him as a mere teller of tales. Even Diodorus Siculus, who is himself a much criticised historian, seems not to have relied on his work any more than a modern Egyptologists will do, unless they have absolutely no other evidence to go on. Herodotus himself said that he just wrote down what he was told! And thank-you that he did. If we endure only as long as our name is remembered then Herodotus has done quite well.

However, as a prime example of the value in checking a legendary source with modern scientific methods we may find no better example than the story that Herodotus gives us for the origins of the Tyrrhenians – or the Etruscans as we would now call them. We encountered Tyrrhenia in the *Timaeus* of Plato – or at least the region of Italy where they later lived –as one of the furthest-east points to which the influence of the Atlantic kings had supposedly reached before their cataclysmic demise. However, megalithic monuments are not found in this region;

these occur only in the south of Italy and Sicily where the Greeks and Carthaginians would later colonise.

Very little is known with certainty about the Etruscans as their language bears no similarity to the Latin and other Indo-European languages around them. Few of their written sources can be translated; and they remain best celebrated for their unique art and architecture. The earliest that we encounter them is around 700 BC, in the area of Tuscany and western Umbria where archaeologists identify the Iron Age Villanovan culture region from about 1100 BC onwards. Their later wars with the Celts and with Rome are historically attested. By about 100 BC they were completely assimilated into the Roman republic and their language was all but extinct.

According to Herodotus the people of Lydia and the islands on the Aegean coast of Turkey had suffered eighteen years of famine. Their king determined that half of the people would have to emigrate and so he made them draw lots to decide who could stay and who should go:

> They passed many countries and finally reached Umbria in the north of Italy, where they settled and still live to this day. Here they changed their name from Lydians to Tyrrhenians, after the king's son Tyrrhenus, who was their leader.[13]

In 2007 two published DNA analyses suggested that Etruscan DNA from their modern descendants in Umbria found matches with that of the Aegean island of Lemnos just as Herodotus suggested.[14] It would seem that at least one ancient migration myth under an eponymous leader might actually be true. However, as ever, specialists find counter evidence and kick back against such conclusions, preferring a more local evolution for the region.[15] They would prefer that 'only some isolates' show descent from Anatolia and are perhaps as old as the Neolithic. So choose which specialist you prefer. Herodotus didn't say that the colonists completely replaced the local population – only that their descendents still lived there. Modern academics, it seems, will go to any lengths to claim that they always know better than the native traditions which ancient people held about themselves.

It has always been uncertain how so much of Roman mythology came to be so similar – and yet so different – to that of the Greeks; and by extension also comparable to the Egyptian. The Romans freely absorbed religious material from all parts of their growing empire. Although the Greeks colonised Sicily and Marseille the religious influence on the Latins must be far older than this; how much of it might have come from the Etruscans has long been debated. For example we find the Etruscan goddess; *Menvra* who clearly equates to the Roman *Minerva*; she in turn is equated with the Greek virgin-warrior goddess *Athene*: the Egyptian *Neit*. Other gods correspond: such as *Nethuns* who was equivalent to Roman *Neptune*: the Greek *Poseidon*; we find *Tina*, the equivalent of Roman *Jupiter* and Greek *Zeus*; *Aril* corresponds to *Atlas*; and in the Etruscan god *Satre* we find the equivalent of Roman *Saturn* and Greek *Cronus*.

The name of the Etruscan god of the underworld is less certain, but we do know of *Charun*, who guarded its entrance with his hammer. The belief in the dark underworld of Pluto and Father Dis that we find in Roman mythology may also have come through from the Etruscan view of death. Like the Egyptians they believed in preservation of the body in sarcophagi in order to be assured of afterlife. Theirs was a far darker vision of the Underworld than even the Hades of the Greeks and certainly much gloomier than the afterlife myths of the Egyptians and the Celts.

It is interesting also that Diodorus Siculus described the expedition of Queen Myrina and her Amazons to the Turkish Aegean coast where they established colonies and religious shrines, *we may suggest*, at some time before 3000 BC. This is the same region from which the Tyrrhenians (Lydians) would later migrate to Italy, taking their religious beliefs and sacrifices with them to the Italian peninsula.

Perhaps we may also see in this Aegean religious heritage the reason why the later Etruscans were so keen to send a colony to the fabled Atlantic island that was the source of tin ore, but were thwarted by the blockade of the Carthaginians. We should note the parallels here with the 'Brutus' legends of British origins. Perhaps they identified in the description of the mythical Atlantic island some memory of the ancient abode of their gods such as we find in degraded form in all the Mediterranean religions.

It seems that when we consider the insular Celtic myths we find recollections of mere ancient kings and heroes; and the land of the dead is remembered as the happy Otherworld: a lost paradise. However when we look at the more distant places that the ancient kings may have conquered or colonised, we find instead mythologies of gods and demi-gods who are worshipped as descendants of unseen deities such as Cronus or Zeus residing in a far-away land. This is rather like we later see the Romano-Britons willing to worship the emperor Claudius as their god, ruling from his citadel in distant Rome – a place that few if any of them would ever see.

We may also observe here the process by which ancient history first degrades to legend when the real historical chronology is lost or was never recorded; it declines further into jumbled myth when it falls into the hands of priests and astrologers, who divert it into religious cults to impress their followers. Truth is a poor passenger in this process, but such are the sources that we have to use to reconstruct a framework for our most ancient history. Surely, anything is better than having to read sagas about the migrations of pottery and flint arrowheads.

The Overlooked Finnic Mythology

In Chapter Five we briefly touched upon the ethnography provided by Herodotus for northern Europe and the visit by Aristeas to the land of the Issedones, which describes the area north of the Black Sea at some time between 1000 and 500 BC. In their distant northern land, we are told, the sea would freeze in winter and to reach them the Scythian traders needed to use interpreters in seven different languages. It was these Issedones who told Aristeas the stories about the Hyperboreans.

Who were these Issedones? In recent times, we find the ancestors of the Estonians and Finns in isolated pockets throughout the vast region of European Russia where their local dialects of the proto-Finnic language still survive. The primary group are the western Baltic Finns and Estonians themselves along with the Karelians, but inland we find other isolated Finnic-speaking groups such as the Volga Finns: the Mordvins and Cheremis. Among the Issedones, we are told, men and women held equal authority, which is a characteristic of Finns and their languages.

It seems that throughout the area of western Russia, where ancient trade in amber took place between the Baltic coast and the Black Sea along the great rivers, there lived a coherent group of proto-Finns whom Herodotus knew as the Issedones. The tribes at this era were constantly under pressure and this region was subsequently infiltrated by the invading Scythians; and they in their turn by the Sarmatians and later eastern invaders. Tacitus tells us about the *Aestii* in his description of the German tribes along the shores of the Baltic Sea some four hundred years later. Unlike the Germans around them they grew corn and traded amber – calling it *glaesum* 'glass'. They are clearly the same people as the earlier Issedones and Essedones. They worshipped the great mother goddess and their language, he says, was unusual, being neither Germanic, nor Celtic – yet we are told by Tacitus that it was similar to one spoken in Britain. This may lead us to the extinct Pictish language of Scotland. The native traditions of the Picts themselves plainly tell us that they were immigrants from the Scythian coast, but Celtic scholars, of course, just dismiss these legends of origin in the time-honoured way that academics always do.[16]

It indeed seems likely that maritime trade and contacts via the Amber Coast and the west had been ongoing since ancient times when Aristeas made his journey to the land of the Issedones. It was they who told him all about the Hyperboreans – but we don't know whether his information was contemporary or if they were feeding him ancient legends. Hecataeus, writing c.350 BC also makes a comment about the 'peculiar' language of the Hyperboreans in an island beyond the Celts. This gives us, if we wish to pursue the possibility, *a further clue* that the pre-Celtic inhabitants of Britain, the people who built the stone circles, were speaking a Finnic language. At precisely what era the southern half of Britain adopted the continental p-Celtic language that would evolve into modern Welsh remains to be established, but these few clues would suggest that it occurred later than, and independent of, the arrival of Celtic speech in Ireland.

The mythology of the Finns are another source of legends against which to cross-check other myths. The only reservation is that it is among the 'youngest' of all the mythologies, having been preserved orally in songs until they were recorded in the mid-nineteenth century. On the positive side, the late conversion of the Finns to Christianity assures us that the pagan myths have suffered less from Christian suppression than the equivalent Celtic stories.

The songs of the Finnish epic *Kalevala* (the land of heroes) describe an underworld called *Manala*, or its equivalent *Tuonela*. The name *maan-ala* means something like 'underground', while Tuonela is the 'land of Tuoni' or 'land of Death'. In Finno-Ugrian myths the Otherworld does not seem to be a paradise like the Celtic Otherworld, nor is it a dark hellish place, it seems rather to be just grim and grey. These names are sometimes used interchangeably in song, but Manala would seem to describe the land of the dead in its entirety, while Tuonela is the island within it. On this isle ruled *Tuoni* and his daughter (or wife?): *Tuonetar*. Finnish shamans, rather like the Greek Odysseus, claimed that they could visit the land of the dead while in a trance, where they would have to deceive the ferryman (or woman) that they were truly dead.

The hero in the Kalevala songs is the old shaman *Väinämöinen* who makes the long journey to Manala to discover the secrets of the dead to use in his magic spells:

> Then to Tuonela he journeyed
> Sought the words in Mana's kingdom
> And with rapid steps he hastened
> Wandered for a week through bushes
> Through Bird-cherry for a second
> And through juniper for the third week
> Straight to Manala's dread island
> And the Gleaming hills of Tuoni
> …
> Väinämöinen old and steadfast
> Raised his voice, and shouted loudly

There by Tuonela's deep river
There in Manala's abysses
Bring a boat, O Tuoni's daughter
Row across, O child of Mana
That the stream I may pass over
And that I may cross the river[17]

However Väinämöinen fails to find the spells that he seeks and is lucky to escape to the land of the living disguised as a snake.

We see again the notion that a ferry boat is needed to cross to the land of the dead, which is a concept common to all the mythologies as far away as Egypt. The 'gleaming' or 'shining' hills on the isle of Tuoni echo again the golden/glass tower of the Celtic myths. Another Kalevala song: *The robbery of the sun and moon*, describes 'the level plains of heaven'.[18]

Many commentators have remarked on the correspondence of the Finnic myths to other European and Egyptian concepts. However, most researchers who have an interest in these subjects will only approach myths as if they were just made-up stories, embodying esoteric religious concepts. It is only when you approach them as degraded history that the similarities with other mythologies become informative.

Coincidences and correspondences between the Finnic and other mythologies are again evident. In all cases the land of the dead and the home of the gods are on an island. A ferry must be used to cross and only the righteous shades may gain entry. In the Egyptian Book of the Dead the city of the gods lies on an island at the mouth of the rivers crossing the Elysian Fields. Comparable names *Manu* and *Tuat* are found as locations within the Egyptian Underworld. The names of *Man* and *Mon* are ancient names of islands in the Irish Sea. The ancient Britons and Gauls had a tradition of ferrying souls to the west of Britain; and Irish and Welsh myths described the happy Otherworld in this same location.

These coincidences are real and they would suggest that a religion from a north European source entered Egypt and the eastern Mediterranean region in ancient times, most probably along with the religion of the goddess Neit. We find the same idea independently in the supporting sources of the other classical authors described above. This explanation may cause a knee-jerk negative reaction from those who slavishly follow the text books, but if they have some other explanation for all the coincidences then they must state their case; as the saying goes: 'If you think your house be stronger than mine then first show me your bricks'.

We may also note that in modern Finnish, the word *nietsyt* means a 'virgin' and *nieti* is the title 'miss' for a young girl. In his Germania, Tacitus tells us about a goddess called *Nerthus* worshipped among the Baltic Germans, whose sanctuary lay somewhere on an island in the sea.[19] We may therefore see an ancient origin for this virgin-warrior goddess among a Finnic-speaking people of Northern Europe, before Indo-European languages intruded into the region. The religion of this goddess and of a Great Mother goddess was carried into the Mediterranean along with a matriarchal or matrilineal culture, by the early megalith builders, where we find it preserved in Libya and Egypt; and in the Neit/Athene/Minerva/Menvra sequence of goddesses. One may find many other potential linguistic connections of this kind, but to follow them all here would be a further digression.

To pursue just one important example, Herodotus also tells us that there were people called *Cynetae*, or *Cynetes* living to the west of the Celts: 'furthest west of all', he says. In modern Finnish the word *kentiä* means 'fields'. So we may put forward that Herodotus was telling us (without knowing so) that, in his time, there were 'fields', or farmers, situated to the west of the Celts. It is yet another of those linguistic coincidences, which probably would not mean much if it stood alone. The name presumably also came via the lost poetry of Aristeas who in turn had it from the Issedones. The reference of Hecataeus to the 'temple of the spheres' in the land of the Hyperboreans, also on an island beyond the Celts, confirms for us that these Cynetae recall the Late Neolithic stone circle builders of Britain: the Hyperboreans. We may still find the ancient name preserved in 'kin' place names throughout Britain: in the county of Kent and the Cantii tribe who lived there; in the names of the river Kennet; and again in the name for the peninsula of Kintyre.

It is another extraordinary coincidence that a sixth century Welsh poem called the *Gododdin* records the battle that took place near Catterick in North Yorkshire, between the Britons of Elmet and the Northumbrian Angles. The poem is an elegy to commemorate the brave North Britons who were killed in the battle, after which the Britons of Elmet lost their independence and their lands were absorbed into Northumbria. The conclusion to the poem describes the people of northern Britain as comprising:

Cynt, a Gwyddil a Phrydin
'The Cynt, the Goidel and the Briton'

The Goidel we know as the Scots/Irish, so here the name Cynt could only refer to the northern Britons of Gododdin or the Picts, who were perhaps ancient kinsmen to the people of Elmet.[20] Modern scholarship will not like this suggestion because they are accustomed to seeing only Celts from sea to sea. The Welsh poem further confirms that this was a name that a people living in Britain applied to themselves. It gives us a further clue, that the Cynetae whom Herodotus described as living 'beyond the Celts' were an independent pre-Celtic nation. We may read the frustration of Herodotus that the Scythians and others whom he asked could not tell him the whereabouts of the Hyperboreans. He does not realise that in telling him of the Cynetae beyond the Celts they had already answered his question.

Traces of the Mesolithic People

Even archaeologists have begun to reassess the capabilities of the Mesolithic people who lived in the west of Europe before the introduction of farming. Perhaps there was more to these early people than just savage hunter-gatherers dwelling in the forests? It is worthwhile to examine on just a few of these clues.

Some specialists have investigated the sea bed offshore of the British coast, but naturally they look first at shallow areas close to the current coastline. In 2015 a team began to investigate the Solent near Bouldnor Cliff between the Isle of

Wight and the south coast of England. In seeking clues to when the site was first submerged they discovered evidence of Mesolithic human activity dating from around 8000 BP. The DNA analysis showed not only the expected pollen of deciduous woodland, but also that of a type of wheat known as einkorn – from around 2,000 years earlier than current evidence of farming onshore and 400 years earlier than the nearby continent. Of course, the cautiously worded study speaks of trading contacts with the continent, rather than suggesting a precocious use of farming in Mesolithic Britain.[21]

Other finds in the north, in the shallow seas surrounding the Western Isles of Scotland have prompted some archaeologists to call for exploration there too, in search of submerged Mesolithic artefacts.[22] This follows the finding of Roman coins and a carved 'chess set' offshore. There has long been folklore that the western isles were formerly linked as a single long island, during the period of human occupation; and that the shallow lochs may have protected sites as much as 9,000 years old. The two specialists who proposed the idea had worked on a similar submerged site in Denmark dating from 6,500 years ago; and suggest that analogous drowning may have taken place west of Scotland. It seems that the 'safe' reason to cite for proposing such submergence is now the Storegga Slide tsunami, which has been advanced as the final act in the severance of the North Sea link. Anything that helps to open-up closed minds is progress.

Further north on Shetland there is evidence from a Neolithic burial cist near Sumburgh Airport. The diet of the earliest inhabitants of Shetland was revealed by a study of their teeth; it showed that the population had temporarily reverted to a diet of sea-food when windblown sand had covered their fields and prohibited their normal agriculture.[23] It would seem that the islands only became viable for occupation after the advent of farming and so as soon as they were able to do so, the people abandoned the fish and sea-food diet and went back to farming. Such evidence probably suggests that people were driven to colonise formerly remote locations simply because they could; and because all the best locations for farming were already taken by a growing population further south. In a later era it was a similar lack of new farmland in Norway that would drive the Vikings across the sea.

Studies in Germany, at the Blätterhölle cave near Hagen, in the Lower Rhine region of Germany, from which the Beaker people of Europe would spread to Britain during the Late Neolithic, have revealed some interesting details about the early farmers.[24] Radiocarbon dates from the cave reveal that it was used as a burial site during two phases; one early in the Mesolithic (9210-8340 cal BC) and again during the Neolithic (3986-2918 cal BC) The specialist's interpretation of the mitochondrial DNA (female line) of the earlier Mesolithic occupants was also found to be present in some of the later Neolithic remains. Furthermore, isotope analysis used to determine the diet of the ancient people revealed that although most were farmers, others were still following a foraging and gathering diet, similar to their Mesolithic forebears. The specialist's conclusion was therefore that the two populations lived alongside each-other; one group farming but the other still following their ancient ways; and that there must have been inter-marriage between the two groups two thousand years after hunter-gatherers were believed to have disappeared. This is another example of how DNA analysis is forcing archaeologists to think differently.

A non-specialist interpretation of such a situation however, has to challenge the polite use of the term 'inter-marriage'. The assumption is that the more culturally advanced farmers took wives from the tribal group, whereas it would be a backward step for a woman from a farming society to join a tribal culture. It is just as likely that such contact was far from voluntary. It may be that the hunter-gatherers were raiding the settled farmers; or alternatively that they were taken as slaves to work the fields. We find this in later Roman times where the German tribes were a regular source of slaves. Such exploitation of tribal people continued into the African slavery of recent colonial times.

Look again at the valuable information that we gain from a reading of Herodotus. Even as late as the first millennium BC the farmers were still recognised as a distinct group living to the west of the Celts in Germany. It prompts us to think again

about the spread of farming from the near-east. It may be that
the coastal cultures had evolved their own pastoral farming and
fruit-growing long before they adopted wheat as a crop; and
that this knowledge advanced around the Atlantic and Baltic
coasts in Mesolithic times as the population grew, leaving the
hunter-gatherers in the central European forests undisturbed
until much later. After all, this is what we observe everywhere
that recent European colonisation has found native societies in
impenetrable and inhospitable jungles. Why make special rules
for the Neolithic colonisation?

And Finally Some Homework...
So much twaddle and scorn has been heaped on the legend of
Atlantis in recent times that it has become unfashionable to
treat it as an authentic source of evidence. In fact, as one may
hope you have discovered from the preceding analysis, there is
much valid material there for study – provided that you treat it
as Egyptian in origin and use it as just *one among many*
comparable sources of myth and legend; and if you accept
what the source actually says rather than what academics and
the twaddle-authors have, in equal measure, done to discredit
it. Once you settle upon a date in the late fourth-millennium
BC for the beginnings of the Egyptian state, then it slots into
its proper place in the pattern of mythological evidence. Note
that word *evidence!* Archaeologists and others who dismiss
and deride ancient legends have for too long neglected to
analyse the software that comes along with the hardware.

Much more investigation is needed to find-out what
happened to the lost Neolithic settlements around the Atlantic
coasts and to understand the true nature of the Mid-Neolithic
Crisis that occurred at the close of the fourth millennium BC.
We can recover history from myths and legends if we approach
the matter in a disciplined scientific way. The DNA evidence
gives us the '*who*'; archaeology tells us the '*where*' and the
'*when*'; and the legends tell us '*what*' they were doing. When
we have all these things for later periods then we quite happily
call it history. So how may we finally sum-up the trail of cross-
disciplinary evidence? In Appendix A is a gist-summary of all
the coincidences that have been followed to reach this point.

When we read the garbled Greek myths of ancient gods and goddesses then we may have a glimpse of the earliest rulers of the Mesolithic-Neolithic transition. We may be looking back some 8,000 years to a time when the first tyrants rose above the level of mere tribal chiefs and became great-kings, defending larger agricultural regions and growing populations.

From the early Neolithic onwards, people of mixed Iberian and North European origin established a common farming culture all along the Atlantic coasts, in the regions where we now find their surviving megaliths and monuments. During the Middle Neolithic a dominant civilization grew that established a colonial empire along the Atlantic coasts and began to expand their influence into the Mediterranean. This was the era of mild and genial climate during the mid-Holocene: the Golden Age.

To the people of the eastern Mediterranean and the east of Europe these distant great-kings living in cities and castles must have seemed almost like gods; and perhaps this was how they actually portrayed themselves. From the trail of evidence we may see that the point of origin for these beliefs lay at the centre of Britain and Ireland. The empire of the Megalith builders was not unlike the later Roman Empire in that it was multicultural. It developed gradually over generations and must have been for most of its existence a peaceful hegemony.

At some time around 3200-3100 BC a series of inexplicable sea level and climate changes occurred, which shattered this early civilisation and caused a severe decline in the population along the Atlantic coasts of Europe, being at its most severe in Britain and Ireland. This crisis most likely had an astronomical cause, which disturbed the Earth's rotation and caused pole tides that inundated parts of the coast. Whether or not you perceive the Atlantis legend or the Elysian Fields as memories of the submerged plain beneath the Irish Sea is a matter that requires investigation. The coastal transgression, the associated climate change and the human recession, was a real event; and the regions lost to the sea were remembered in the Celtic mythology.

The years between 3100 and 2900 BC were a hiatus in the development of Atlantic Europe. The population along the coast and islands was severely reduced and this allowed people of Steppe ancestry to emerge from the forests; and for colonists from the Mediterranean region to venture west. For a time, as we see in the Irish invasion myths, the survivors tried to keep them out, or they might permit some controlled immigration. Again, we may draw the analogy with recent European colonisation of the Americas or Australia. After a few hundred years the new settlers outnumbered the aboriginal inhabitants and had absorbed what remained of their culture.

For those of us of a certain generation; almost everything that we were told about prehistory, by the experts, up to the 1970s has proven to be wrong. The advent of radiocarbon dating along with tree-ring calibration demolished most of the certainties that had grown-up about ancient chronology – but a few old ideas persisted, particularly in the derivation of European languages. The new DNA evidence is now challenging older dogma about the Celts and the origin of the people of the Britain, Ireland and Atlantic Europe. Perhaps in this new climate, archaeologists and other academics may at last begin to show some respect for the ancient myths and legends.

Notes and References

[1] Olalde, Inigo, et al, 2018, *The Beaker Phenomenon and the Transformation of North West Europe*, Nature 555, 190-196 (5 March 2018)

[2] The precise wording is: 'Our results imply a minimum of $93\pm2\%$ local population turnover by the Middle Bronze Age'.

[3] Strabo, *Geography*, 4.5.4.

[4] I am now unable to trace this report, which I read, most likely, in the Yorkshire Post or Harrogate News in the early 2000s. One of my own employees actually attended the auditions.

[5] Gerald of Wales, *The Description of Wales*, Book II, Ch.r 1.

[6] BBC News report, 19 July 2011

[7] BBC News report 19 June 2012

[8] Yorkshire Post Wednesday 31 October 2012 *"It's official – DNA tests show Yorkshire people really are a different breed"*
[9] Geoffrey of Monmouth, *History of the Kings of Britain*, i, II; Lewis Thorpe translation.
[10] Bear in mind that we don't know with certainty either the date or the reality of the Trojan War – it is another example of unproven scholarly opinion accumulated over time to become pseudo-fact.
[11] Arrian, I, IV; quoting here Alexander's biographer, Ptolemy Soter.
[12] See the comments of Douglas Hyde, *The Literary History of Ireland*, 1899, p 7.
[13] Herodotus I, 93-96.
[14] *https://www.britishmuseum.org/pdf/14%20Perkins-pp.pdf*
[15] Am J Phys Anthropol. 2013 Sep; 152(1):11-8. doi: 10.1002/ajpa.22319. Epub 2013 Jul 30.
[16] I have investigated this relationship more thoroughly in *Picts and Ancient Britons*. It stems from a symposium of Celtic scholars in 1950 which produced a report called: *The Problem of the Picts*.
[17] Runo VXI, *Väinämöinen in Tuonela*, 150-166; translation of W.F. Kirby, 1907.
[18] Runo XLVII, line 181
[19] Tacitus, Germania, 40
[20] The region of Gododdin lay between modern Edinburgh and the Roman wall.
[21] Smith, Oliver, et al; *Sedimentary DNA from a submerged site reveals wheat in the British Isles 8,000 years ago*, Science, 347,6225, 27, Feb 2015, pp 998-1001.
[22] BBC News 13 July 2011 citing Dr Jonathan Benjamin and Dr Andrew Bicket
[23] Melton, N.D. & Montgomery, J. (2009) *Combined isotope analyses of Early Neolithic individuals from a burial cist at Sumburgh, Shetland*, Archaeology Scotland.
[24] Bollongino. R. et al, (2013) *2000 Years of Parallel Societies in Stone Age Central Europe*, Science, 362, 25 October 2013, pp 429-481.

Appendix A

A Summary of Coincidences

Time

- King lists show that the earliest period of the Egyptian state begins around 3100 BC not at 10 500 BC.

- The correct era to seek evidence of a catastrophic event is contemporary with Predynastic and First Dynasty Egypt (late fourth millennium BC).

- Plato's description should be treated as an Egyptian chronicle recorded by Solon c.590 BC rather than as a fiction of Plato around 350 BC.

- Solon's description shows correspondences with the earliest Libyan and Carthaginian sources as cited by Diodorus Siculus.

- Both Plato and Diodorus describe cultural and religious colonisation emanating from a civilisation along the Atlantic coast.

- Solon's description of Atlantis shows similarities to the Elysian Fields depicted in the Egyptian Book of the Dead.

- Crantor is likely to have seen maps of the Elysian Fields that matched Solon's description.

- First Dynasty traditions of the cult of Neit at Saïs were revived during the Twenty-sixth Dynasty, contemporary with the visit of Solon to Egypt.

- The sources that Solon saw at Saïs were probably lost during or after the Persian Invasion of 525 BC.

- The worldwide pattern of sea level change shows a correspondence in alternate quarter-spheres since 3000 BC.

- Climate changes in Europe show transition from the Atlantic to Sub-Boreal pollen zone c.3000 BC.

- Emergence of the Nile Delta and transition to a full desert climate in the Sahara date from c.3000 BC onwards.

- Tree ring low growth event at 3199 BC.

- Greenland ice-core sulphate spike 3250 BC.

- Era of the Mayan calendar at 3113 BC.

- Era of the Indian calendar (Kaliyuga) at 3102 BC.

- The culture of the megalith builders flourished at this period all along the Atlantic coasts and into the Mediterranean.

Manner

- Plato and Diodorus Siculus describe submergence in the west, but only earthquakes in the eastern Mediterranean.

- A pattern of sea-level change in alternate quarter-spheres would be consistent with adjustment after a pole-shift.

- Comparison with the Lisbon earthquake shows that the 3100 BC event must have been an order of magnitude greater than the largest earthquake and tsunami events.

- Submergence in a single day and night is consistent with a pole tide event

- Pole tides can only occur along with a free wobble lasting around 20 years (the Chandler wobble).

- Permanent submergence suggests a small pole shift (approximately a quarter degree of latitude).

- A Core-wobble must also occur along with the Chandler Wobble.

- The Core-wobble takes 2000-2500 years to decay to rest.

- The Core-wobble causes seven-year rhythms in the climate and more extreme seasons.

- Plato's Timaeus mentions a 'winding' of the rotation axis, which may describe a vestige of the Core-wobble.

Place

- Other Mediterranean mythologies also remember the Elysian Fields.

- The Elysian Fields were believed to be in 'the west', or situated on an island.

- In the era of Solon (c.600 BC) the Greeks had forgotten the geography of the Atlantic coast.

- Knowledge of the Atlantic coast and islands was preserved in Phoenician and Carthaginian sources.

- The Carthaginians remembered a mythical 'paradise' island in the Atlantic.

- The Carthaginians regularly traded tin ore with Britain and perpetuated legends about an island 'opposite' Spain.

- An extant real island, at the centre of three large islands, remembered a former time when it was part of an Atlantic Empire (Marcellus).

- Britain and Ireland are the only real large islands in the Atlantic that would fit the description given by Plato and the Libyan sources.

- Britain and Ireland lie 'opposite' the north coast of Iberia.

- The Carthaginian route to the mythical island was the same as that to Britain and Ireland.

- British and Gaulish myths describe the ferrying of souls to a 'land of the dead' situated west of Britain (Procopius).

- Irish mythology remembers an Otherworld situated in the Irish Sea near the Isle of Man.

- Irish mythology remembers the earliest inhabitants as Fomorians or 'people from under the sea'.

- Welsh mythology remembers an Otherworld called Annwn adjacent to Wales.

- Welsh mythology describes a plain lying off the north coast of Wales.

- Welsh myths recall a time when it was possible to walk from Britain to Ireland.

- Irish legend remembers a land bridge from Ireland to Scotland (Giant's Causeway).

- Mammals and reptiles were able to reach both Ireland and the Isle of Man since the Ice Age.

- Submerged forests around the British west coast give dates c.3100 BC.

- Welsh myths remember sunken cities around the British coast.

- Anglesey was the sacred island of the Druids.

- Druidism was invented in Britain and there was a Druid college situated somewhere in Britain.

- The 'central island' that preserved knowledge of the Atlantic culture was most likely Anglesey and its Druids.

- Welsh and Irish legends recall a golden or 'glass' tower submerged in the sea.

- The eastern Irish Sea is a submerged 3:2 rectangular plain with a mountain at the centre (Isle of Man).

- On the Irish Sea floor is a smaller submerged 3:2 rectangular plain between Isle of Man and North Wales.

- Submerged river channels can be traced on the Irish Sea floor, which discharged to the ancient sea near Anglesey.

- The topography of the eastern Irish Sea floor matches the Egyptian drawings of the Elysian Fields.

- Sonar surveys show patterned ground on the Irish Sea floor.

- Monuments and pottery suggest a common culture all around the Irish Sea during the Middle Neolithic before 3100 BC.

- Irish legends speak of a 200 year period during which Ireland was depopulated.

- There was a decline in farming and peat or forests grew on deserted fields in the years after 3100 BC.

- DNA studies show that over 90% of the British Neolithic population was replaced by new immigrants after about 3000 BC

Appendix B

Pole Shifts and Pole Tides

The physics of the Earth's rotation is not simple to understand and cannot be made so. However, to appreciate how pole shifts might have affected the climate and world sea-level during human prehistory requires at least a basic grasp of the physics in order to understand what is possible. So, with apologies to the geophysicists the following is a summary using as little jargon as possible, so that non-specialists who might wish to study it further will have a point of entry.

Physical geographers in the nineteenth century, who would come to be called geophysicists, began to investigate how the Earth would behave if the daily rotation were disturbed. This came about, in part, because some scientists suggested that a *pole tide* might be an explanation for the Biblical flood of Noah. So what is a pole tide? Firstly, more explanation is needed. The principle argument that led the nineteenth century scientists to dismiss the possibility of pole-shifts was that the energy requirement to change the *obliquity*: the 'tilt' of the axis in space, by any significant amount would be so huge that it would destroy us all – we would not be here to discuss it.

A change of obliquity is not the same as a geographical pole shift. It is important to distinguish between an external force that could affect the attitude of the axis *in space* and a displacement of the figure axis *within the body of the Earth* that changes the location of the poles. A pole shift only requires an internal change to the shape of the planet. Since the Earth's figure is oblate due to the daily rotation this would also modify the sea level all over the world. Small pole shifts do in fact occur after every earthquake because they alter the figure of the solid Earth. For example, the magnitude-9 Sumatra earthquake of 2004, in addition to producing a tsunami wave, also caused a pole shift of about 2.5cm and shortened the day by milliseconds. This gives you a sense of the vast energy that must be released in order to cause a really significant pole-shift of the order suggested by our myths and legends; but the same maths and physics would apply. However, before the axis can migrate it first has to wobble.

Two modes of wobble occur because the Earth has a fluid nickel-iron core. These are termed *free wobble* because no external force needs to act and they decay exponentially to rest. The first is a wobble of the crust and mantle (the shell) alone, the second can only occur if the shell began to rotate about a different axis from that of the core. It is important to appreciate that the two modes must occur together and both have a spatial and a body component. The axis 'nods' in a motion called a *nutation*.

The first mode is known as the *Chandler wobble*, after its nineteenth century discoverer. It is a nutation of period 435 days – mainly within the body of the Earth and a corresponding but negligible motion in space. Estimates of the decay or relaxation period say 15-30 years; but probably about 20 years before it would become unnoticeable to a casual observer. Visualise it as a wobble of the shell-only, in which the core plays no part. Geophysicists regularly measure these small polar motions in the order of arc-seconds.

The second mode is known as the *nearly-diurnal wobble* because it is a retrograde circuit of just under a (sidereal) day. This occurs on top of a retrograde motion in space that is about 460 times larger. It has a period estimated as 460 days and a lifetime of perhaps 2000-5000 years. As the rotation today is stable this motion is barely measurable and so its precise characteristics remain theoretical. Some sources estimate the period as 434-444 days and the lifetime somewhat shorter.

For decades the nearly-diurnal wobble was neglected by geophysicists and not well understood. In the 1980s some geophysicists even claimed to have detected the tiny motion – until it was pointed-out that it can only occur on top of the wobble *in space* that should be 460 times larger and (if it existed) should be noticed by every amateur astronomer. This larger motion didn't even have a name up to that time. In my own earlier book I was forced to give it a name: 'core-mantle precession' in order to talk about it without resort to equations. However it is now generally discussed in geophysical papers as the *core wobble* or the *free core nutation*.

An important principle to appreciate is that after the effect of the Chandler wobble has decayed away the core wobble would persist. We don't notice the core wobble after an earthquake because its amplitude is too small to measure. However, after a significant pole-shift event the core wobble should persist for perhaps 2500 years. It would trigger seven-year rhythms in the climate, so if it has ever happened in the past then geophysicists should be able to detect it in the ancient climate data.

For most of the twentieth century the terminology used by geophysicists was confusing – and not only for the non-specialist! Since the 1990s there has been a tightening of terms in the published papers, so anyone wishing to research further should look in their search-engines for certain terminology. The body-wobble is best known as the Chandler Wobble. The tendency now is to use *nutation* for the spatial component and *wobble* for the body component. The space-based component of the Chandler motion is termed the Diurnal Nutation or DN. Also look for *Free Core Nutation* (FCN), also known as the *Nearly Diurnal Free Wobble* (NDFW). Look also for *polar motion* and *Earth-fixed reference frame* (ECEF).

So, to return to the subject of a pole tide; the Chandler wobble occurs when the Earth changes its shape for whatever reason. The axis of symmetry – the figure axis – jumps to a new position, as do the points where it cuts the Earth's surface – the rotational poles; but the Earth continues to rotate about the same axis in space. The figure axis then begins to spiral around the rotation axis until, after 20 years or so the wobble is damped and they again reunite. The period of each circuit is about 14.7 months. Throughout this motion the solid surface remains aligned with the figure axis, whereas the oceans must immediately conform to the rotation axis. This is the pole-tide. Depending where you are on the Earth's surface, the sea would either rise to overflow the shore, or it would retreat to expose the sea-bed. These effects would progress around the planet with the period of the Chandler wobble. After the wobble is damped the world-wide sea level would be left symmetrical about the new figure axis. The accompanying nearly-diurnal wobble (which would be ongoing for much longer) should not produce a significant pole tide.

A pole shift would produce a pattern of permanent sea-level change in alternate quarter-spheres. However the Earth's crust is flexible at the plate boundaries and transform faults, so can gradually adjust its shape, leaving just a few places on the Earth's surface where the coastal displacement would be permanent.

The standard glacio-eustatic model of sea level change neglects the possibility of small pole shifts in recent Earth history.

Some Useful References:

Lambeck, K. (1988), *Geophysical Geodesy - The Slow Deformations of the Earth*, Clarendon Press, Oxford.

Toomre, A. (1974). *On the 'nearly diurnal wobble' of the Earth*, Geophys. J. 38, 335-48.

Gross, Richard S (2000). *"The Excitation of the Chandler Wobble"'*. Geophysical Research Letters. 27 (15): 2329–2332.

"Earth's Chandler Wobble Changed Dramatically in 2005". TechnologyReview.com. MIT Technology Review. 2009.

INDEX

213

INDEX